The Dreams of Reason

THE DREAM

COLUMBIA UNIVERSITY PRESS

New York and London

OF REASON

Science and Utopias

BY

RENÉ DUBOS

To
Adeline De Bloedt Dubos
and
Nora Wills Porter

The George B. Pegram Lectures

Opportunities for reflective appraisal of the import of science on society by the very architects of current events and trends should be cultivated whenever possible. To this end the George B. Pegram Lectureship was established by the Trustees of Associated Universities, Inc. Residence at Brookhaven National Laboratory gives the lecturer opportunities for formal and informal contacts with the staff, and provides a period of freedom from other duties.

As the second lecturer in the series, Dr. René Dubos has elected to reexamine the influence of Bacon on the modern structure and development of science, the limits to which scientific inquiry and the scientist are subject, as well as the dilemmas which arise in the pursuit of medical research and alleviation of man's ills. Dr. Dubos, member of the Rockefeller Institute, is well known both in his specific fields of microbiology and experimental pathology and as an author. His latest volume is *The Mirage of Health,* published last year. His books also include the third edition of *Bacterial and Mycotic Infections of Man* (1958); *Biochemical Determinants of Microbial Disease* (1954); *The White Plague—Tuberculosis, Man and Society* (1952); *Louis Pasteur—Free Lance of Science* (1950); and *The Bacterial Cell* (1945).

The lectureship was named to honor George Braxton Pegram, one of the most influential scientists of the nuclear age. He was Professor of Physics, Dean, and Vice President of Columbia University. He was instrumental in seeing that the government was aware of the potentialities of nuclear energy in the defense of the country. In 1946 he headed the Initiatory University Group which proposed that a regional center for research in the nuclear sciences be established in the New York area and thus played a key role in the establishment of Associated Universities, Inc., and the founding of Brookhaven National Laboratory. He received many awards and honorary degrees, the last of which was the Karl Taylor Compton gold medal for distinguished service in physics. George B. Pegram's lucid mind and gentle ways will be long remembered by those who knew him. This series in his honor has been established to further his conviction that the results of science can be made to serve the needs and hopes of mankind.

Gerhart Friedlander
Samuel A. Goudsmit
Leland J. Haworth
Daniel E. Koshland, Jr.
James S. Robertson
R. Christian Anderson, Chairman

Acknowledgments

The author gratefully acknowledges permission to quote from the following volumes: *A Diderot Pictorial Encyclopedia of Trades and Industry* by Denis Diderot (Dover Publications); *The Firmament of Time* by Loren Eiseley (Atheneum Publishers); *The Inward Vision* by Paul Klee (Harry N. Abrams); *A Land* by Jacquetta Hawkes (Random House); *A Modern Utopia* by H. G. Wells (Chapman & Hall); *The Pasteur Fermentation Centennial 1857–1957* (Chas. Pfizer & Co.); *The Revolt of the Masses* by José Ortega y Gasset (W. W. Norton & Company). The caricature of H. G. Wells reproduced on page 51 appears in *Lions and Lambs* by Low, with interpretations by "Lynx" (Harcourt, Brace & Company).

Contents

1. Background and Prejudices   1

2. Salomon's House and the Baconian World   12

3. Visionaries and the Era of Fulfillment   40

4. Medical Utopias   63

5. Illusions of Understanding   99

6. The Dehumanization of the Scientist   129

7. The Humanness of Science   157

Illustrations

Goya's "Dreams of Reason"   13

Francis Bacon   20

Raphael's "School of Athens"   31

A plate from Diderot's Encyclopedia   36

Sir Thomas More   45

H. G. Wells   51

'Tisick, Colic, and Gout   68

Rembrandt's portrait of a physician   97

René Descartes   104

Lichens   112

Charles Darwin   117

Michael Faraday   140

Louis Pasteur and his wife   143

Thomas Edison   147

Bernard de Fontenelle   151

Thomas Huxley   154

The birds are to be envied:

They avoid

Thinking about the trees and the roots.

Agile, self contented, all day long they swing

And sing, perched on ultimate end.

PAUL KLEE

The Inward Vision

The Dreams of Reason

1

Background and Prejudices

Within a month's time during the fall of 1960 I passed
from the National Laboratory in Brookhaven to the hill
town of Assisi in Italy. The change was not traumatic
despite the contrast, because both places bring man closer
to fundamentals. In the humble monastic garden of Assisi
where St. Francis wrote the "Hymn of the Sun" almost
eight centuries ago, I recaptured in thought the peculiar
warmth in the voice of a Brookhaven physicist explaining
the mysteries of the proton. St. Francis apprehended and
served the creation through love. And the words carved
on the entrance to the huge cosmotron in Brookhaven
seem to me congenial to his teachings: "Dedicated to the
Enlightenment and Service of Mankind." I shall have
nothing more to say concerning Francis of Assisi, but
what I have said is sufficient to warn the reader that I
find it difficult to dissociate science from the rest of human experience, a confession that may help him to understand the mood in which this book has been written.

This book is based on the George B. Pegram Lectures that I delivered at the Brookhaven National Laboratory in October, 1960. The circumstances which brought me to Brookhaven are of some interest because they give to the lectures a significance that far transcends their merit and content. The Brookhaven National Laboratory is a research center devoted to studies in nuclear physics and in the biological effects of radiation. Unfortunately for me, these most exciting and mysterious aspects of modern science are far removed from the areas of medical biology in which lies my own professional specialization. The invitation to be the Pegram Lecturer took me, therefore, by surprise because I felt completely unqualified to speak before the Brookhaven scientific staff. I was assured, however, that my assignment was not to discuss technical aspects of physics, chemistry, biology, or medicine, but rather to present my views on science as a part of the social structure and of the humanistic tradition. It is the very nature of this assignment that I regard as worth emphasizing, because it throws into relief an aspect of scientific life which is commonly ignored.

Everybody knows that science is the most effective instrument for the creation of wealth and power in the modern world. Most laymen also acknowledge that scientists as a group exhibit great dedication to their tasks, even though their financial rewards are rarely commensurate with their efforts. But while scientists may be admired and even envied, it must be admitted that they are not loved. They are often accused of being narrow in their interests, insensitive to the emotional and artistic aspects of life, unconcerned with the social consequences of their actions—in brief, of lacking human warmth and understanding. I have tried in this book to present what seems to me a truer picture of the scientific community. The decision of the Brookhaven National Laboratory to devote the Pegram Lecture Series to the social and humanistic implications of science constitutes a good introduction to this topic because it illustrates an attitude which, contrary to general belief, has long been prevalent among scientists. As I evoke the two weeks spent in Brookhaven among scientists working at the frontiers of knowledge and understanding, what I recall is not narrowmindedness, arrogance of intellect, or indifference to human feelings, but instead their eagerness to share in all the experiences, hopes, and responsibilities of mankind. Nowhere have I appreciated more fully that scientists consider their calling not merely a practical trade from which they derive satisfaction, but truly a vocation dedicated to the needs of the human soul and to the fulfillment of man's destiny.

The theme around which my lectures were organized emerged in part from the historical accident that the year 1961 marks the four hundredth anniversary of Francis Bacon's birth. When I received the invitation to deliver the Pegram Lectures, it occurred to me that this type of lectureship, and indeed the very existence of such institutions as the Brookhaven Laboratory where they would be delivered, can be traced to the social influence of Bacon's writings, and to his utopian description of Salomon's House in *The New Atlantis*. Hence the association between science and utopias which came to my mind even before I knew precisely what the content of the lectures would be.

It is likely that Bacon's quadricentennial will be celebrated in many learned books and conferences. Furthermore, there is no need of historical knowledge to predict that his role as a prophet of modern science will be highlighted during the celebrations. Some biographers will certainly take the view that it was Bacon who publicized most effectively the role of the inductive method in experimental science and who convinced the world that scientific knowledge is power. Other historians are likely to point out that Bacon was merely the voice of his age and that scientists long before him had practiced what he put into such sonorous words. Thus, a debate will probably take place between those who trace the activities of scientists to the formulation of large problems by philosophers and those who believe that modern social philosophy has emerged out of the findings of science. I thought that a contribution to this debate would constitute a suitable topic for the Pegram Lectures, since these have been established "to provide a forum for discussing the broad implications of science in our times."

More specifically, I shall attempt to show that the illusions, aspirations, and whims of mankind, even more than its physical needs, influence profoundly the beliefs and activities of scientists. While I do not question, of course, that scientists are objective in the actual performance of their experiments, it seems apparent to me that they plan their studies on the basis of large philosophical and social assumptions. For example, they may believe that pure knowledge is more important than power, or take it for granted that the production of more wealth and better drugs is the only effective approach to happiness and health. They may assume on faith that life is the expression of some divine vital spirit, or accept—*also* on faith—that living processes are but the expression of the activities of a kind of chemical molecule which happens to be fashionable at the time. Some form of belief is necessary for action, but it is dangerous not to be aware of the underlying assumptions which condition one's thoughts and behavior. What I shall try to do here is to bring to light some of the unproven assumptions which have influenced the theories and practical applications of science.

Most of the examples that I shall use in the following essays will be taken from biology, and a few from chemistry and physics. This selection reflects my personal experience, and does not imply that I separate in my mind the natural from the social sciences. In fact, I find it extremely difficult, if not impossible, to define science, to delineate the traits that differentiate it from other human activities. Everyone agrees, of course, that theoretical physics and genetics are within the scientific domain and that lyric poetry and symphonic music are not. But what about psychology, sociology, or history? The most I can say is that, granted much uncertainty as to the limits of science, a few convictions concerning its character are held by all scientists and most philosophers. The quest for the real, the verifiability of assertions, clarity in the definition of the terms employed, consistency in the affirmations stated in these terms, and humble respect for all the discoverable facts relevant to the problem under study are all fundamental to the practice of science—constitute, indeed, its very bedrock. Yet, while there is no doubt that science is primarily concerned with real, verifiable events, it is also true that many scientists spend their most pleasant professional hours, and often their most creative, in dreamlands unencumbered with realities and beyond the reach of verifiability. Indeed, it seems to me that the activities of even the most objective and practical of experimenters are conditioned not only by tools, techniques, and logical concepts, but also, and perhaps even more, by conceptual views which transcend factual knowledge. In a large measure, then, this book will be concerned with the interplay between two aspects of scientific life: on the one hand, the factual and verifiable components and, on the other, the imaginative and emotional determinants.

I know that my analysis of the human factors in science cannot be thorough, and I suspect that it will not be objective. The difficulties will originate not only from my lack of knowledge, but even more from the fact that my attitude toward science has been indelibly conditioned by what I read and heard during my student days. It is rather easy for me to identify these early influences because I can still quote from memory a number of statements that struck me at the time. In order to help the reader apply the necessary correcting factors to the opinions that I shall express, it may be appropriate that I briefly describe the views of science which I gained as a young student and which are still vivid in my consciousness. As to subconscious determinants of my attitude, I am convinced of their existence but shall not try to identify them.

I was born and educated in France and read little outside the writings of French authors until I was an adult. Fortunately, the French cultural tradition is so rich and diversified that it gave me a chance to see science through several colors of the spectrum. But I now realize that many wave lengths were entirely filtered out and, furthermore, that I was insensitive to some of those which came through.

As probably is true of many boys all over the world, my first contact with science was made through Jules Verne. I am sure that his stories could not instill valid knowledge or critical judgment or scientific spirit in anyone, but they can certainly foster a taste for the unknown and a desire for adventure. This is not without importance, because the longing for the uncommon and for the unexpected is a powerful motivating force in many scientists—indeed, perhaps the only one in some of them. I shall have occasion later to point out that curiosity and the search for adventure are important factors in scientific life.

Very early, of course, I learned about Pasteur and about the applications of his discoveries in industrial processes and in medicine. Through Pasteur's example the scientist appeared to me as the benefactor of mankind. Needless to say, I heard of many other scientific achievements which were of great practical importance and changed the face of the world; but I did not appreciate the dynamic character of science, the fact that new discoveries are constantly changing technology and adding to its potentialities for new processes and new products. Science then appeared to me more as a body of principles and of facts to be learned than as a method for increasing endlessly the total amount of knowledge and of technological application. Whether my teachers and the books I read failed to emphasize the creative dynamism of science or whether I was not responsive to its meaning and implications is a question I cannot answer. But the fact itself is certain, and it has been of importance in my subsequent attitude. The utilitarian aspects of science have never loomed large in my vision, and, while I recognize their importance in the abstract, my intellectual and emotional involvement in them has always been lukewarm.

In contrast, other aspects of science were constantly emphasized in courses of history, philosophy, and literature, as well as in the endless conversations among students.

These discussions dealt with the role of the scientific point of view and of scientific facts not only in man's understanding of the universe and of his own nature, but also in the selection and interpretation of subjects by artists and writers. It is of interest in this regard that the textbook of French literature which I used in school analyzed the writings of great French scientists such as the physiologist Claude Bernard, the microbiologist Louis Pasteur, the chemist Marcelin Berthelot, and the mathematician Henri Poincaré, and illustrated them with large photographs of these authors. As Ernest Renan has said, "Everything becomes great literature when done with talent." It goes without saying that there were conflicting views among the teachers who tried to define for us the place of science in civilization, and I believe that I was even then very confused as to my own opinion concerning the relative importance of literary and scientific studies.

Most of us had read *L'Avenir de la science*, the famous book in which the philosopher Ernest Renan announced to the world that science was the new faith which had replaced in him the Catholicism of his youth. In impressive language Renan asserted that science alone could solve the enigma of the world and "reveal to man in a final form the real nature of things." Moreover, according to him, scientific facts were soon to provide a new and richer kind of material for poetical inspiration. "Although the tales of fiction have been assumed so far to be essential to poetry," he wrote, "the true wonders of nature will provide a far more sublime subject once they have been unveiled in all their splendor; they will be the source of a poetry that will be reality itself, that will be at the same time science and poetry."

Those of us who did not appreciate or trust Ernest Renan's oratory could quote statements just as sonorous made by famous scientists of the same period. Was it not Pasteur who spoke lovingly of the experimental method as "admirable and sovereign" and said of scientific research: "The charm of our studies, the enchantment of science . . . consists in the fact that everywhere and always, we can give the justification of our principles and the proof of our discoveries"? Even more positive and forceful was the famous chemist Marcelin Berthelot, who asserted in 1901 that "Science is today in a position to claim the leadership of societies, not only with regard to material questions, but also to intellectual and moral problems. . . . It is science that will provide the truly human basis of morals and politics in the future." And elsewhere Marcelin Berthelot did not hesitate to affirm his confidence that "We are justified in pretending, without going beyond legitimate assumptions, that we can conceive and create the general types of all possible forms of life . . . that we can synthesize all the substances which have been developed since the beginning of time, under the same conditions and according to the same laws used by Nature to produce them."

There is no need to illustrate by further quotations this unlimited confidence in the power of science which was so general a faith among scientists during the nineteenth and early twentieth centuries, and which amounted at times to narrow-minded intellectual arrogance. I shall have occasion in the following essays to deal at some length with the reaction against science generated in many intellectual groups by this overconfidence. Suffice it to mention here the one example which brought me into contact with the antiscience movement during my student days.

Ernest Renan was only twenty-five years old in 1848 when he wrote his book *L'Avenir de la science,* but he did not publish it until 1890. His faith in science had obviously weakened in the interval, as can be seen from some of the statements in the preface to his book: "While I still believe that it is only through science that the human condition can be improved, I no longer believe that the solution of the problem is as near us as I had once hoped. . . . I am afraid that the main contribution of science will be to deliver us from superstitions rather than to reveal the ultimate truth." Renan's skeptical words in 1890 did not correspond merely to an intellectual pose; they reflected an attitude that was becoming common in some intellectual groups throughout Europe at the turn of the century. The theme *"la faillite de la science"* (the bankruptcy of science) was popular among philosophers and littérateurs, and it was heard even in some scientific circles. To illustrate the mood of skepticism to which I was exposed, I might quote a sentence from Anatole France which must have impressed me greatly at the time because I still remember it today, forty years after reading it. "I despise science, my boy, for having loved it too much," Anatole France said through his mouthpiece Jérome Coignard, "like those disenchanted voluptuaries who reproach women for not having given them the satisfaction that they had expected from love."

In this personal narrative of my early student years I have tried to evoke the atmosphere of a lively but not too sophisticated school environment in France before, during, and shortly after World War I. I now realize that what I took so seriously then was the very superficial and diluted backwash of much deeper controversies that were going on all over the learned world. Everywhere and at different levels of sophistication there was concern about the nature of scientific truth and the influence of science on civilization. An English student with a comparable background would have been exposed to H. G. Wells's optimism based on a boyish faith in science, but also to G. B. Shaw's scornful remark that "Science is always wrong. It solves problems only to replace them by others." A German student would have been steeped in the arrogant self-confidence of thorough Teutonic scholarship, but also in the romantic and diffuse airings of *Naturphilosophie*. And even in America the practical attitude symbolized by pragmatism was counterbalanced by the fundamentalist creed that led not so long ago to the antievolutionary "monkey trial" in Tennessee.

While no one can escape the determining effects of his genetic make-up, it is almost as difficult to escape the conditioning influence of early cultural experiences. I have presented here a narrative of some of my early memories relevant to science in the hope that this will help the reader compensate for my lack of objectivity. In any case, the Pegram Lectures are intended to be a forum, and subsequent lecturers will introduce other facts and points of view in discussing the relation of science to modern society.

2

Salomon's House
and the Baconian World

Today, as every day, I have heard of ugly congested cities with polluted atmosphere; of planes loaded with youths colliding in mid-air; of overpopulated continents and starving populations; of mechanized, regimented, and dehumanized life; of brainwashing and nuclear warfare. As a member of the scientific community, I am awed by the thought that these social nightmares are to a large extent the products of industrial civilization—born out of science. And there comes to mind an etching by Goya, the central plate of his *Caprichos* series. It shows a man sprawled across his desk, daydreaming or asleep, his head on his arms. Bats, owls, and a witch's cat surround him as in a nightmare. On the side of the desk are inscribed the words *"El sueño de la razon produce monstruos."*

I had assumed that this caption meant that the *sleep* of reason produces monsters. For, indeed, errors and superstitions readily take over and generate loathsome creatures when reason is asleep. It is more likely, however,

This etching, known as *"The Dreams [or Sleep] of Reason"*
from the inscription on the desk,
is one in Francisco Goya's Los Caprichos *series (1796).*
Goya's own caption for the etching can be translated:
"Imagination deserted by reason creates monstrosities.
United with reason, imagination gives birth
to great marvels and true art."

that the caption refers not to sleep, but to the undisciplined *dreams* of reason. To Aldous Huxley the etching means that "reason may intoxicate itself as it did during the French Revolution." At that time reason engaged in daydreams of endless and inevitable progress, of utopias to be reached through the easy road of political liberty and more advanced technology.

Like Goya—for he himself is the dreamer sprawled over the desk in the etching—I recall that men of good will and of reason in all ages have believed that the problems of the world would soon be solved by science, the very science which instead has engendered some of the monsters that threaten us today. One of these men of reason, among the finest of eighteenth-century dreamers, was M. J. A. N. Caritat, Marquis de Condorcet, author of the famous *Sketch for a Historical Picture of the Progress of the Human Mind.* "I will show," Condorcet asserted in his essay, "that Nature has set no limit to the perfecting of the human faculties, that the perfectibility of man is truly infinite; that the progress of this perfectibility, henceforth independent of any power that might wish to arrest it, has no limit other than the duration of the globe on which Nature has placed us. . . . The time will therefore come when the sun will shine only on free men who know no other master but their reason."

Although Condorcet died a victim of the French Revolution, he apparently never lost his faith in the future of mankind. But one cannot help wondering how he would react today, less than two centuries after his death, when so many thoughtful men question the very survival of civilized life. In contrast to the optimistic mood of the eighteenth century and the hopes of endless perfectibility of man through science, this is what an

American scientist—Loren Eiseley in *The Firmament of Time*—has to say of twentieth-century America, the land where scientific civilization has reached its highest level:

Modern man is being swept along in a stream of things, giving rise to other things, at such a pace that no substantial ethic, no inward stability, has been achieved. Such stability as survives, such human courtesies as remain, are the remnants of an older Christian order. Daily they are attenuated. ... We have re-entered nature, not like a Greek shepherd on a hillside hearing joyfully the returning pipes of Pan, but rather as an evil and precocious animal who slinks home in the night with a few stolen powers. The serenity of the gods is not disturbed. They know well on whose head the final lightning will fall.

Progress secularized, progress which pursues only the next invention, progress which pulls thought out of the mind and replaces it with idle slogans, is not progress at all. It is a beckoning mirage in a desert over which stagger the generations of men.

These devastating statements seem to echo the lament of another contemporary American scholar. Said Joseph Wood Krutch in *Human Nature and the Human Condition:* "What some of us tend to call 'the human being' first came into easily recognizable existence about the year 475 B.C. and began to disappear about seventy-five years ago."

The humanization of mankind was the flowering of reason. As reason falls asleep or becomes intoxicated, monsters take command of civilization and man loses his humanity, even though he may gain wealth and power. Many are those who believe that an uncontrolled appetite for the products of modern technology is intoxicating human reason; let us, then, examine how science, so long an adornment of the mind, has now come to be valued chiefly for the creation of wealth and power and to be regarded by many as an instrument of evil.

Technology is, of course, very ancient—indeed, as ancient as mankind. Many of the most important inventions can be traced far back in history, some of them to prehistoric times. From the practices of farming and of the potter's trade to the development of gunpowder and of the magnetic compass, the list of the technological achievements whose originators will always remain unknown is very long. Likewise, many abstract scientific concepts have come to us from our Oriental and Greek heritage. Mathematical formulas and astronomical observations, ideas concerning the structure of matter and the evolutionary changes that occur in living things, and so many other problems of theoretical science had been formulated and some of them converted into practical applications long before the modern era. But while science has been practiced by man since the beginning of time, there does not seem to have existed in the distant past the concept that it could be disciplined and organized into a systematic body of operations applicable to all human problems. What is really peculiar to the modern world is the belief that scientific knowledge can be used at will by man to master and exploit nature for his own ends.

The change in point of view which led to the modern attitude toward science seems to have taken place sometime around the fourteenth century in western Europe. Since there is no reason to assume that the human brain was less well endowed in antiquity than it was during the Renaissance when the modern attitude developed, historians have found it necessary to appeal to all sorts of extraneous reasons to account for the surprising fact that science was practiced for so long in such an amateurish manner. One reason frequently given is that the widespread practice of slavery prevented the Orientals and the Greeks from engaging in ordinary manual labor, and thus deprived them of the inspiration that can be derived from intimate contact with concrete problems. Slave societies, it is claimed, do not have the urge to create effort-saving processes and equipment. It has been pointed out also that the form of Christian doctrine which prevailed in Europe until the fourteenth century weakened human interest in the affairs of this world by placing so much emphasis on the afterlife. Thus, two aspects of the social changes that occurred in the post-medieval era would have contributed to making conditions more favorable for an awakening of the modern scientific attitude: on the one hand, the emergence of the bourgeois and artisan classes with experience in the concrete affairs of the world; on the other, the more material and worldly attitude that began to develop throughout Europe at the time of the Renaissance.

The historical basis of these explanations is not entirely convincing. For example, recent studies have established that the number of slaves in Greece was far smaller than is commonly assumed, and that much of the physical work of everyday life was carried out by free men. As to medieval civilization, there is no reason to believe that its Christian faith, however deep, interfered seriously with the enjoyment of material pleasures. It seems safe to assume, therefore, that factors other than slavery and preoccupation with an afterlife were of importance in the intellectual revolution associated with the Renaissance.

There is no doubt, for example, that machines and

gadgets—such as mechanical clocks—were popular during late medieval times, and their operation may have encouraged the thought that the workings of nature could be explained by simple mechanical forces.

As the Florentine painters became more concerned with the naturalistic aspect of art, their probing into the structure of the human body and of other living things led to a more objective and concrete view of the world. The discovery that the time-honored teachings of Galen on human anatomy and physiology were often erroneous, especially with regard to the functions of the heart, shook confidence in traditional knowledge. Progressively the revolt against medieval scholasticism gave greater freedom to the mind for appealing to forms of explanation and of authority other than religious dogma and Aristotelianism. Concrete experience appeared as valid and meaningful as the evidence derived from the Gospels and from Greek and Roman writers; and soon it proved more fruitful in both intellectual and material satisfactions than religious and scholastic orthodoxy.

Incomplete and unconvincing as these explanations are, they help to recapture the human atmosphere in which modern science emerged. The ferment which agitated the men of the Renaissance led them to undertake scientific studies as a way to manifest their spirit of independence and their desire to do things, whatever these things might eventually turn out to be. Thus, the new science was an expression of will to power, an instrument forged by the human mind before man knew what he would use it for. In any case, all accounts of the beginnings of modern science point to the fact that scientific activities are intimately interwoven with the social fabric. We shall consider only a limited segment of this complex interrelationship—namely, the influence that philosophical thought has exerted on the performance of the scientific community.

As mentioned above, it is clear that science has been coexistent with mankind, but there is no doubt, on the other hand, that scientific discoveries were made at an accelerated rate after 1400. In fact, one of the most articulate anticipations of the role that science was eventually to play in the world occurred even earlier, in the writings of the monk Roger Bacon during the thirteenth century. In addition to doing some scientific work of his own, this worthy predecessor of the more famous Francis Bacon emphasized that much useful knowledge could be obtained from the experience of practical tradespeople. Furthermore, he was imaginative enough to predict the invention of flying machines, of power-driven ships, and of other unheard-of contrivances that man might wish to use. However, despite all the premonitory signs detectable during the Middle Ages—whether actual discoveries or optimistic writings about science—the scientific revolution which began yielding its fruits during the seventeenth century can be traced in large part to the writings of one person: Francis Bacon. His place in the history of science is unique because his influence was exerted through words rather than deeds. He did not add to knowledge, but became the prophet of scientific civilization.

It would be useless to tell once more the life history of Francis Bacon, Baron Verulam and Viscount St. Albans, who was born in London on January 12, 1561, and died there on April 9, 1626. But it is of direct relevance to our theme that, in addition to his interest in

Francis Bacon (1561-1626) is usually shown
in his court attire, with all the dignity
attached to the function of Grand Chancellor of England.
Here he appears in his later years as the scholar,
exiled from political office.
This portrait was the frontispiece
in Of the Advancement and Proficience of Learning (*1640*).

science, he was heavily involved in political activities of the most practical sort, an expert on legal matters, a writer of essays, and such a master of the English language that there are still some today who believe that he was the true author of Shakespeare's writings. His sense of practicalities and his extraordinary genius for literary expression were the main factors in determining his role in the subsequent history of science.

Francis Bacon, "wisest, brightest and meanest of mankind," as Pope characterized him, had such a full public life that a biography of him published in 1940 by W. M. Cunningham had the title *The Tragedy of Francis Bacon—Prince of England*. Despite all his worldly activities, however, Bacon pretended that his professional and business life was not particularly important in his eyes and that his contemplative life was what he prized above all. When still a young man he had written in a letter that his ambition was "to take all of knowledge for his province," and he apparently tried hard to live up to this ideal. Throughout his crowded legal and parlimentary career he kept at work upon a grandiose scheme for a *Great Instauration* or total renovation of the sciences. In 1605 he published *The Advancement of Learning*, a classification and critical survey of all existing knowledge, and in 1620 his greatest work, the *Novum Organum*, an exposition of the new experimental method.

Bacon recognized, of course, that important contributions to science had been made in the past, and that still more were being made by his contemporaries. But he believed that on the whole the business of scientific scholarship had been conducted in a wasteful manner, contributing little to factual knowledge and even less to the improvement of the human condition. It will be best from now on to present Bacon's thoughts on the whys and hows of the pursuit of knowledge in his own words because it was more the brilliance of his language than the originality of his ideas which influenced subsequent generations of scientists and scholars.

Bacon based his scientific philosophy on the Christian dogma of Original Sin. While man had lost at the same time the state of innocence and his dominion over the external world, Bacon believed that these losses could be repaired to some extent on earth. Man could recover Adam's original state of happiness on the one hand by religious faith, on the other by the cultivation of science. In this light, "knowledge is not to be sought either for pleasure of the mind, or for contention, or for superiority to others, or for profit, or fame, or power, or any of these inferior things; but for the benefit and use of life. . . ." "The true and lawful goal of the sciences is none other than this: that human life be endowed with new discoveries and powers. . . ." "As in religion we are warned to show our faith by works, so in philosophy by the same rule the system should be judged of by its fruits, and pronounced frivolous if it be barren; more especially if, in place of fruits of grape and olive, it bear thorns and briars of dispute and contention."

To sum up, then: "Knowledge, that tendeth but to satisfaction is but as a courtesan, which is for pleasure, and not for fruit or generation."

How can mankind use science to recapture the dominion over nature that was lost by the Fall? First we must shake off, according to Bacon, our intellectual bondage to the ancients, for they have so completely dominated our thoughts heretofore as to paralyze action. There is no doubt, of course, that "the ancients proved themselves in everything that turns on wit and abstract meditation, wonderful men. . . ." "Unfortunately," Bacon says, "that wisdom which we have derived principally from the Greeks is but like the boyhood of knowledge, and has the characteristic property of boys: it can talk, but it cannot generate; for it is fruitful of controversies but barren of works."

Moreover, what has come down to us from the past, in particular the teachings of Plato and Aristotle, is the least valuable part of ancient knowledge and wisdom, because it is the most superficial. "Time is like a river, which has brought down to us things light and puffed up, while those which are weighty and solid have sunk. . . ." "When on the inundation of barbarians into the Roman Empire human learning had suffered shipwreck, then the systems of Aristotle and Plato, like planks of lighter and less solid material, floated on the waves of time, and were preserved."

The greatest evil that we have inherited from the ancients is that their approach to knowledge is more conducive to talk than to action. "They had made the quiescent principles, *wherefrom*, and not the moving principles, *whereby*, things are produced, the object of their contemplation and inquiry. The former tend to discourse, the latter to works."

Our most important task, then, is to discover how things work rather than to answer questions about their origin. In the past, knowledge has been derived chiefly from the mere observation of things and events, and from speculative thought about the scanty facts thus acquired, but this is a very ineffective method, and one which does not permit us to go far. "As in former ages when men sailed only by observation of the stars, they could indeed coast along the shores of old continents or cross a few small and mediterranean seas; but before the ocean could be traversed and the new world discovered, the use of the mariner's needle, as a more faithful and certain guide, had to be found out: in like manner the discoveries which have been hitherto made in the arts and sciences are such as might be made by practice, meditation, observation, argumentation—for they lay near to the senses, and immediately beneath common notions; but before we can reach the remoter and more hidden parts of nature, it is necessary that a more perfect use and application of the human mind and intellect be introduced."

In order to progress we have to develop precise techniques of experimentation, because "the nature of things betrays itself more readily under the vexations of art than in its natural freedom. . . ." "The subtlety of experiments is far greater than that of the sense itself, even when assisted by exquisite instruments; such experiments, I mean, as are skillfully and artificially devised for the express purpose of determining the point in question. To the immediate and proper perception of the sense therefore I do not give much weight; but I contrive that the office of the sense shall be only to judge of the experiment and that the experiment itself shall judge of the thing."

Bacon knew, of course, that experimentation had been practiced long before his time. But "the manner of making experiments which men now use is blind and stupid. And therefore, wandering and straying as they do with no settled course, and taking counsel only from things as they fall out, they fetch a wide circuit and meet with many matters but make little progress . . . make their trials carelessly, and as it were in play." What was needed, therefore, was a discipline of experimentation based entirely on strict application of the inductive method. "Whereas in the past the proceeding has been to fly at once from the sense and particulars, up to the most general propositions, as certain fixed poles for the argument to turn upon, and from these to derive the rest by middle terms . . . my plan is to proceed regularly and gradually from one axiom to another, so that the most general are not reached till the last; but then when you do come to them you find them . . . such as lie at the heart and marrow of things."

Thus, Bacon was convinced that man can achieve new things and improve the world only if he gives up haphazard observation and experimentation. He must formulate far-reaching goals and organize efforts in a more subtle and systematic way. Bacon recognized that not all experiments could be expected to lead immediately to practical results. Indeed, "scientists should be willing to carry out a variety of experiments, which are of no use in themselves, but simply serve to discover causes and axioms; which I call *experimenta lucifera,* experiments of light to distinguish them from those which I call *fructifera,* experiment of fruit.

"Now experiments of this kind have one admirable property and condition: they never miss or fail. For since they are applied, not for the purpose of producing any particular effect, but only of discovering the natural cause of some effect, they answer the end equally well which ever way they turn out; for they settle the question."

Bacon's emphasis on what he called *experimenta lucifera* (experiments of light, or *expériences pour voir,* as Claude Bernard was to say two centuries later), devised specifically to throw light on obscure problems before attempting to solve them, shows a truly sophisticated understanding of the experimental method. It is clear that he did not advocate a formula of science limited to the mere heaping up of experimental data. Nowhere do the subtleness of his mind and the richness of Elizabethan language appear more brilliantly than in these famous sentences: "The men of experiment are like the ant; they only collect and use; the reasoners resemble spiders, who make cobwebs out of their own substance. But the bee takes a middle course, it gathers its material from the flowers of the garden and of the field, but transforms and digests it by a power of its own." Clearly to be effective the scientist had to resemble the bee, industrious but also imaginatively selective, in order to be truly creative.

Bacon also realized that even the most objective experimenter with a well-defined goal could not be successful if his angle of vision was too narrow: "No one successfully investigates the nature of a thing in the thing itself; the inquiry must be enlarged, so as to become more general." But, suspicious of the tendency of human nature to engage in the artificial kind of idle thought which leads nowhere, he wanted all scientific activities to be built on the bedrock of concrete problems. "Although the roads to human power and to human knowledge lie close together, and are nearly the same, nevertheless on account of the pernicious and inveterate habit of dwelling on abstractions, it is safer to begin and raise the sciences from those foundations which have relation to practice, and to let the active part itself be as the seal which prints and determines the contemplative counterpart."

All in all, he trusted the practical sense of mankind

more than its intellectual pretenses. "Let no man look for much progress in the sciences—especially in the practical part of them—unless natural philosophy be carried on and applied to particular sciences, and particular sciences be carried back again to natural philosophy. . . ." "Brutes by their natural instinct have produced many discoveries, whereas men by discussion and the conclusions of reason have given birth to few or none."

Lest this statement be construed as evidence that Bacon regarded science merely as an instrument for crass material ends, without intellectual quality, it is important to point out that he knew well that the surest approach to practical applications is through the painstaking and slow process of acquiring theoretical knowledge. "Though it be true that I am principally in pursuit of works and the active department of the sciences, yet I wait for harvest-time, and do not attempt to mow the moss or to reap the green corn. For I well know that axioms once rightly discovered will carry whole troops of works along with them; and produce them, not here and there one, but in clusters."

Moreover, man can recover his dominion over nature only through understanding the secrets of nature, above and beyond immediate practical ends. In final analysis, "Truth, therefore, and utility are here the very same things: and works themselves are of greater value as pledges of truth than as contributing to the comforts of life."

It must suffice to quote one more example to illustrate Bacon's wide-ranging awareness of the bearing that science has on human life. He suspected that knowledge, when placed in unworthy hands, might come to be used for destructive and cruel purposes, but this did not dismay him. "If the debasement of arts and sciences to purposes of wickedness, luxury, and the like, be made a ground of objection, let no one be moved thereby. For the same may be said of all earthly goods; of wit, courage, strength, beauty, wealth, light itself."

So great was the role of science as envisioned by Bacon that he wanted to see practical steps taken to help scientists fulfill their social responsibilities. He was shocked to note that "Natural philosophy . . . has scarcely ever possessed, especially in these later times, a disengaged and whole man (unless it were some monk studying in his cell, or some gentleman in his country house), but that it has been made merely a passage and bridge to something else."

Probably as a form of protest against the social neglect of science, he depicted in the last of his writings, *The New Atlantis,* a utopian society guided by a community of scholars who devoted themselves entirely to scientific research, to the organization of knowledge, and to the pursuit of wisdom. In the words of their leader, "The end of our foundation is the knowledge of causes, and secret motions of things; and the enlarging of the bounds of human empire, to the effecting of all things possible." Thus did Francis Bacon, Lord of Verulam, Grand Chancellor of England, symbolize in Salomon's House his ideal of the scientific way of life, and in the utopian New Atlantis his concept of a society intelligently ruled by scientific philosophers.

Despite his early claim that he wanted to take all learning for his province, it is certain that, even for his time and by any standard, Bacon himself was not much of a scientist. True enough, he applied his system of inductive science to an analysis of the nature of heat and he derived by theoretical analysis conclusions which were compatible with the dynamic theory, but his mode of reasoning and of handling information seems very strange indeed in the light of modern experimental science—especially in comparison with the marvelous scientific achievements of some of his predecessors or contemporaries such as Galileo Galilei, William Gilbert, and William Harvey. His involvement in matters of law and of state did not give him much chance to prove his worth as an experimenter, and, ironically enough, the last experiment that he carried out was responsible for his death. While traveling in March, 1626, he stepped out of his coach, procured a chicken, had it killed, and helped to stuff it with snow, in order to see whether the flesh would remain fresh longer if kept at a low temperature. He contracted some respiratory disease as a result of exposure to the cold, and died of it shortly after.

From the vantage point of modern knowledge, it is easy to criticize Bacon's concepts of the methods of scientific research. As we now know, very few important discoveries—if any—have been made by applying strictly the pure inductive method that he advocated with so much conviction. Indeed, we shall notice in another chapter that productive scientists approach their problems by all sorts of methods—inductive as well as deductive, rational as well as empirical—and this was already true during Bacon's time. It is entertaining in this regard to note the evaluation of the Baconian scientific philosophy expressed by his own physician—the great physiologist William Harvey, who opened the modern era in medicine by discovering the circulation of the blood. Harvey was, of course, familiar with Bacon's writings and he knew the tables of instruction which his famous patient regarded as foolproof guides for making discoveries. According to John Aubrey, Harvey esteemed Bacon for his wit and style "but would not allow him to be a great Philosopher. He writes Philosophy like a Lord Chancellor." Harvey's remark betrays the scornful amusement and slight irritation that practicing scientists always feel toward those who express opinions on scientific matters from secondhand knowledge without really knowing the secrets of the trade. It must be emphasized, however, that Bacon was well aware of his limitations as a scientist and that he considered that his main role had been to act as a gadfly. In his words, he "rang the bells which called the wits together."

Having perceived in the official and lay world around him a lack of enthusiasm for science and skepticism arising from failure to appreciate its potentialities, he affirmed that "the greatest obstacle to the progress of science and to the undertaking of new tasks and provinces therein, is found in this—that men despair and think things impossible." In contrast, he considered that his method constituted only a beginning, because "the art of discovery may advance as discoveries advance." He made no claim of being superior to his predecessors. "The comparison I challenge," he wrote, "is not of wits or faculties, but of ways and methods, and the part I take upon myself is not that of a judge, but of a guide."

Bacon's significance in history is thus to have blown the clarion call which awakened Europeans to the fact that science could completely transform society. He was not the first, of course, to have recognized that science could be applied to the welfare of man. Aristotle had stated in *Mechanics* that "vanquished by nature, we be-

Raphael's "The School of Athens,"
now in the Vatican, is shown here
in an engraving by Volpato.
Plato and Aristotle occupy the center of the stage,
surrounded by Pythagoras, Heraclitus, Diogenes,
Archimedes, etc., in characteristic postures.
Raphael's symbolic representation
of science as philosophy stands in sharp contrast
to the illustrations in Diderot's Encyclopedia,
where science appears as technology.

come masters through techniques." And knowledge had always been applied to the development of tools and weapons for use in peace and in war. Leonardo da Vinci had boasted of that kind of scientific skill in his famous letter to Ludovico Sforza, offering to reveal his technical secrets and to prove to the Duke by tests that he was capable of doing things thought impossible by others.

What Bacon advocated, however, was a very different approach. He emphasized that in the past the applications of knowledge to the practical affairs of man had not been systematic but instead in the nature of stunts, accidental in occurrence, and kept secret if possible. In contrast, he preached a general method by which problems could be solved at will, thus permitting a progressive and continuously increasing mastery over nature through the systematic and uninterrupted pursuit of knowledge. Moreover, he urged that these applications of science be made available to mankind at large, instead of being the privilege of a few fortunates. In this light, he truly appears as one of the prophets of the modern age. A recent biography by Benjamin Farrington (1949) is appropriately entitled *Francis Bacon, Philosopher of Industrial Science.*

Bacon was a man of the world, full of experience and wisdom. Being aware of the complexities of the universe and of the many difficulties that stood in the way of applying knowledge for the welfare of mankind, he realized that the goal he had in mind was distant and that its attainment would require the efforts not of one man but of many, not one lifetime but generations of men working with a common purpose. This aspect of Bacon's teaching had a profound influence on his and on the following generations, by introducing the concept of continued sci-

entific progress through scientific cooperation. His utopian New Atlantis constitutes a preview of a society built
on this concept. In Salomon's House the scientists who
were the wise men of this ideal state were organized in
nine groups, each with a special function, according to
the principle of division of labor. Technological and
physical laboratories, as well as agricultural experiment
stations, were at their disposal; and elaborate records
were kept to preserve and to communicate knowledge.
Some scientists planned the work, others did the experiments, still others recorded and organized information.
"Lastly," said the leader of the House, "we have three
that raise the former discoveries by experiments into
greater observations, axioms, and aphorisms. These we
call Interpreters of Nature."

In general, scientists in the Europe of Bacon's time
worked as isolated individuals because organized groups,
let alone institutes for research, did not exist. In other
words, the ideas of scientific organization incorporated
in Salomon's House were then completely utopian; yet
many of them came into being within a very few decades
after Bacon in the form of scientific academies and
periodicals.

This is not the place to list the many institutions that
sprang up in Europe and also in America during the seventeenth and eighteenth centuries, directly or indirectly
as a result of Bacon's influence. One of them, however,
deserves mention here because it was the first to be created and is still today the most famous—namely, the
Royal Society of London, which was officially founded in
1660, three centuries ago. The Royal Society had begun
in the 1640s as an informal gathering of a few scientists—

an invisible college, as they called themselves. Wren, the celebrated architect of St. Paul's Cathedral, was one of its first members, as was the chemist Robert Boyle. In a letter to his French tutor the young Boyle stated that he had been studying "natural philosophy, the mechanics, and husbandry, according to the principles of our new philosophical college, *that values no knowledge, but as it hath a tendency to use.*" (Italics mine.)

After receiving approval from Charles II in 1660, the Royal Society prospered to such an extent that a history of it was compiled five years later by its secretary, Bishop Thomas Sprat. In this history, published in 1667, Sprat acknowledged that "some of Bacon's writings" gave a better account of the purposes of the Society than anything he could compose. And in many of its characteristics, indeed, the Royal Society of early days appears as an embodiment of Bacon's teachings. Men of all professions were admitted—"students, soldiers, shopkeepers, farmers, courtiers, and sailors; all mutually assisting each other." The work of the Society was not literary or esoteric but rather "painful digging and toiling in nature." It was concerned with practical trades and industries as much as with abstract scientific pursuits.

The progress of science through cooperation was the aim of the new scientific societies. As stated by Oldenburg in the introduction to the first issue of the *Philosophical Transactions* (1666), edited by the Royal Society, the new periodical was being published

to the end that such Productions being clearly and truly communicated, desires after solid and useful knowledge may be further entertained . . . and those, addicted to . . . such matters, may be invited and encouraged to search, try and find out new things, impart their knowledge to one another, and contribute what they can to the Grand design of improving Natural knowledge. . . .

The French Academy of Sciences was another scientific institution founded to put into practice the concepts of progressive and organized knowledge popularized by Bacon. Colbert, the minister of Louis XIV, had created the Academy in 1671 and endowed it with funds for supporting the Academicians and paying for their instruments and experiments. In some ways the French Academy came closer than the Royal Society to Bacon's utopian dreams, as it was officially charged with the responsibility for studying technical problems of national interest, carrying out experimental work in various fields, and disseminating the results of scientific inquiry. Only during recent decades has Bacon's vision come even closer to reality in the form of the academies of the U.S.S.R. and their scientific institutes.

Another spectacular manifestation of Bacon's lasting prestige was the respect with which the philosophers of the Enlightenment mentioned his name. Several French translations of his writings were published during the eighteenth century, and it is certain that they influenced the character of the most famous literary scientific document of the time: *The Encyclopedia, an Analytical Dictionary of the Sciences, Arts, and Trades.* The first of the twenty volumes of this monumental undertaking was published in 1751, with a preface by Diderot and D'Alembert which is justly famous. In it the editors acknowledged their intellectual indebtedness to Descartes, Newton, and Locke, and especially to Bacon.

At the Head of these illustrious Heroes we deservedly place the immortal Francis Bacon, Lord High Chancellor of England; whose works, though justly esteemed, are too little

A shop for threading screws is shown
in this plate from Denis Diderot's
Pictorial Encyclopedia of Trades and Industry (1763).
In Figure 1 the workman is tracing the pattern
of a thread on a steel rod.
The machine in Figure 3 is a threader,
as on a much larger scale is that of Figure 4.
The latter is turned by the wheel (5)
which is cammed to reverse the rotation
while moving back and forth
the width of the thread with each revolution.

known, and deserve Perusal more than Praise. To consider the just and extensive Views of this prodigious Man; the Multiplicity of his Objects; the Strength of his Style; his sublime Imagery; and extreme Exactness; we are tempted to esteem him the greatest, the most universal and most eloquent of all Philosophers. . . . It is to this great Author we are chiefly indebted for our Encyclopaedic Plan.

From Bacon the Encyclopedists had learned to respect the practical arts as much as theoretical knowledge. In their words, "Should not the inventors of the spring, the chain, and repeating parts of a watch, be equally esteemed with those who have successfully studied to perfect algebra?" This point of view was voiced still more eloquently by Diderot in his article on "Art" in the *Encyclopedia·*

Let us at last give the artisans their due. The liberal arts have adequately sung their own praises; they must now use their remaining voice to celebrate the mechanical arts. It is for the liberal arts to lift the mechanical arts from the contempt in which prejudice has for so long held them, and it is for the patronage of kings to draw them from the poverty in which they still languish. Artisans have believed themselves contemptible because people have looked down on them; let us teach them to have a better opinion of themselves; that is the only way to obtain more nearly perfect results from them. We need a man to rise up in the academies and go down to the workshops and gather material about to be set out in a book which will persuade artisans to read, philosophers to think on useful lines, and the great to make at least some worthwhile use of their authority and their wealth.

Although the examples quoted in the preceding pages illustrate Bacon's seminal influence on the scientific philosophy of the modern era, they do not prove that he was completely original in his views. Indeed, authors before him had expressed more or less clearly a belief in progress through science. It is easy to show, furthermore, that Bacon was somewhat naïve and ill informed, even for his time, with regard to the way science grows and true scientists work. The community of scholars that he pictured in Salomon's House had many shortcomings and in the long run would probably have become intellectually sterile. But it is not on these inadequacies that Bacon must be judged. His greatness is to be found in his eloquent and passionate affirmation that science would become a great social force. He appears as the first statesman whose aim it was to organize human life in terms of a master plan framed by scientific thought. Unknowingly, William Harvey had characterized Bacon's greatness in saying that he wrote on science "like a Lord Chancellor"—in other words, like a great statesman of science.

In *The Advancement of Learning* Bacon advocated the methods of natural philosophy for the improvement of health and for improving civil conduct. In *The New Atlantis* he drew a sketch of a commonwealth making use of technology in every department. He was truly the prophet of things to come, and hence it is not surprising that his ideas have found wide acceptance in the socialist states. Karl Marx was merely copying him when he wrote, "Hitherto, philosophers have sought to understand the world. Henceforth they must seek to change it."

It was the richness and convincing beauty of Bacon's language that made the world at large take notice of scientific knowledge as an instrument of power and of social growth, thus launching us on the road that we are still traveling today. No one questions any longer the fact that science is increasing the dominion of man over nature. This does not mean, of course, that man will recover through scientific technology the happiness that

Adam knew before the Fall, as Bacon hoped. But Bacon certainly contributed to the modern world its most characteristic aspect and its most lasting illusion when he created his utopia of happiness based on application of scientific knowledge.

3

Visionaries

and the Era of Fulfillment

History is replete with anecdotes and bons mots relating to statesmen, soldiers, artists, philosophers, and most other types of notables; but even a well-informed man finds it difficult to enliven talk with quotations from scientists. The reason might be that scientists as a group lack wit or are inarticulate, but there is no evidence that this is true; more plausibly it could be suggested that the number of scientists has been so very small until the present century that the statistical chance of their having proffered quotable remarks is thereby greatly reduced; finally, there is the certain fact that historians have never been much interested in scientists and for this reason have failed to record their words of wisdom. Yet, anyone who has read in the history of science and of its makers can recall statements attributed to scientists which deserve to survive either for their significance or for their pungency. Indeed, a collection of such sayings might throw interesting light on certain human aspects of scientific history, and especially on the changes in the public's attitude toward science.

While Benjamin Franklin was ambassador at the French court, he witnessed some of the first balloon ascents. In reply to skeptics who queried what use a balloon might ever have, Franklin is asserted to have replied, "What good is a newborn baby?" This story has been often quoted—for example, by Pasteur, who used it as an argument to convince his students of the importance of scientific studies during the lecture which opened his new course in chemistry at Lille in 1854. A story in the same vein is told of Michael Faraday. Shortly after he had discovered electromagnetic induction, Faraday reportedly was visited by an important personage in his laboratory at the Royal Institution. He demonstrated the phenomenon to his visitor, who was unimpressed and inquired, "What is the good of this discovery?" Faraday is alleged to have replied, "Someday, sir, you will collect taxes from it."

Whether these stories are authentic or not is of limited historical interest. What *is* important is that they imply an awareness of the power and social significance of scientific experimentation. As we have seen, this awareness can be traced in large part to Francis Bacon, but it did not reach the general public until the nineteenth century.

The answer "What good is a newborn baby?" clearly symbolizes the faith that scientific discoveries are not an end unto themselves, that they hold in potential many further developments, not all of which can be readily predicted because so much depends on the future circumstances that will bring them to maturity. And the suggestion that taxes will someday be collected from a new scientific fact is but a more concrete and worldly statement of the same faith. It expresses the belief that almost any scientific discovery will eventually be converted into some process or product which society can use, and for which it is willing to pay. Apparently these truths were still new in the early part of the nineteenth century. In fact, they have not yet been completely assimilated even in the most industrialized countries—witness the occasion a few years ago when a Secretary of Defense in President Eisenhower's cabinet scoffed at basic science as dealing with such "useless" subjects as "why grass is green."

Despite the fact that he had been president of a huge industrial corporation, this important personage did not fully realize that the multifarious industries over which he had presided were much more the outgrowth of theoretical scientific research than of his own skill as a financier and an administrator. But the explosion of scorn and ridicule that greeted his remark was evidence than an increasing number of persons are now aware that science has become the real source of power in our society. Furthermore, it is now widely recognized that the most unexpected results of scientific research often turn into its most exciting fruits—those bred without too much concern as to the use of a newborn baby.

The ignorance of science and the naïveté revealed by the remark of the Secretary of Defense has an interest of its own, for it points to the fact that today the social role of science is still more a matter of faith than of precise understanding. Indeed, the problems involved in the relation of science to society are on the whole so new that our knowledge of them is certainly incomplete and probably erroneous in part. For this reason, it may be permissible to engage in a few farfetched speculations, linking scientific progress to the formulation of utopias by moralists, philosophers, and sociologists.

In the matter-of-fact world that we know, the words "imagination" and "imagining" have lost much of their quality and have acquired instead a somewhat pejorative meaning, at least in the scientific community. They have come to imply a distorted awareness of reality, often coupled with a lack of intellectual discipline. And yet these words have their origin in one of the most creative characteristics of the human mind—indeed, one of the very few traits differentiating man from higher animals. To "imagine" clearly means to create an image—more precisely, to select from the countless and amorphous facts and events which impinge upon us a few that each individual can organize into a definite pattern which is meaningful to him. This is what Shelley had in mind when he wrote in *A Defence of Poetry,* "We want the creative faculty to imagine that which we know." To imagine is an act which gives human beings the chance to engage in something akin to creation.

During many thousands of years, men have used the elements of the real world with which they came into contact to imagine—that is, to create in their minds—other worlds more reasonable, more generous, and more interesting. These acts of imagination have had an enormous influence on history, as great as or perhaps greater than the effects of newly developed processes and tools. For it is certain that in many cases new processes and tools have found their place in civilization only when they could be used to actualize, to bring into being, the imaginary worlds first conceived in the abstract by the human mind. In this light, imagination has been one of the most creative forces of civilized life, because it has provided the molds which mankind has used to shape the crude facts of reality into significant structures.

Since Plato formulated a plan for an ideal society in his *Republic*, many would-be philosophers have described imaginary states in which the political, social, and economic schemes were assumed to be conducive to health and happiness. Of all the imaginary worlds, past and present, none has achieved greater, wider, and more lasting fame than Sir Thomas More's *Utopia*. First published in Latin in 1516, often reprinted, first translated into English in 1551, retranslated into English in 1684, put into almost all European languages and, in the twentieth century, into several Asiatic ones, More's *Utopia* is obviously universal in its appeal and ubiquitous in its influence. It is acclaimed by Christians and Communists alike, by Protestants and Catholics, by progressives and reactionaries—indeed, by men of all schools of thought, though for widely different reasons. The facts that More was recently canonized by the Roman Catholic Church and that in the same decade his *Utopia* was adopted as a textbook in Soviet Russia indicate the difficulties involved in assessing the man and his book. (See figure 5)

For most modern readers it is difficult to find in *Utopia* anything specific to account for its phenomenal and lasting fame. Like many authors before and after him, More used the artifice of an imaginary world to criticize the institutions of the real world he knew—in this case, Tudor England—and to describe, in contrast, his ideal of a society ruled with common sense and fairness. Although first published some 450 years ago, More's *Utopia* deals with topics which are still very much in the public mind today,

Holbein's portrait of Sir Thomas More (1478-1535)
is reproduced by permission of the Frick Collection
in New York. The reproduction does not do justice
to the piercing quality of More's eyes
in the original, but conveys his forcefulness and wisdom.
It was probably More's prestige,
rather than the content of his famous book,
which gave to the word "utopia"
the lasting fame that it enjoys in the Western World.

such as penology, juvenile delinquency, education, and social responsibility. The state of Utopia is governed by laws which are the expression of unaided human reason, without divine guidance; eugenics is practiced with the assistance of premarital tests, divorce is easy, and euthanasia is permissible—even encouraged and facilitated in certain special situations. In other words, More's essay reads like a commentary on today's newspaper; its perennial and universal timeliness has certainly contributed to making it a world classic.

It is probable that in More's mind the word "utopia" implied nothing more than its literal meaning—"not place," a place that does not exist—and had no greater significance than the word "nowhere" in the account of another imaginary world that William Morris published three centuries later under the title *News from Nowhere*. It was the success of More's book among so many different groups of people that made the word "utopia" evolve in two directions more or less independent one from the other. "Utopia" now refers to an ideal state, but also designates something considered rather unreasonable because almost certainly impossible to realize. Thus, the word symbolizes both the wishful thinking of the romantic idealist and the resistance to change of the skeptical conservative.

As mentioned above, utopias have been imagined in every generation and in every country. Most of them are inspired by hopes of political, social, and economic reforms, as were Plato's *Republic* and More's *Utopia*. Bacon's *New Atlantis* is one of the very few utopias in which the problem of making society wise, kind, and prosperous is entrusted to scientists.

Although the writers on utopias have on the whole kept their creations safely sheltered within the covers of books, their followers have not been so cautious. It is claimed that two hundred or more utopias were put into actual practice during the nineteenth century, chiefly in America. Examples are the so-called Love Colony; Brook Farm, the colony of intellectuals; and, most famous of all, Robert Owen's New Harmony in Indiana. The rapid collapse of all of them stands as a warning to those who want to put idealistic philosophies to the acid test of economic realities. Whereas the concept of utopia is remarkably viable as long as it remains in the abstract, all utopias in the flesh have soon perished. In part, their rapid disappearance from the real world resulted from failure to pro vide man with the earthly goods that he requires. Furthermore, all utopian dreams of human harmony were soon dispelled in the heat of human conflicts and rivalries under practical conditions. Utopias invariably bring out the traits of human nature—and there are many obvious ones—that stand in the way of unselfish and stable social relationships.

Accustomed as he was to comfort and luxury, Bacon pretended that scientific research in his New Atlantis could adorn the material necessities of existence with the amenities essential for civilized life; the scientists in Salomon's House enjoyed beautiful clothing and spacious marble halls. Indeed it is likely that no utopia can succeed if it concerns itself only with the biological requirements of life. Man does not live by bread alone, and his parabiological requirements created by social forces have become as essential to him as the needs imposed by his animal nature. Whenever shortages of bread or brandy develop, jealousies and conflicts arise, emotional tensions mount, regulations become necessary, and liberties are lost.

To provide adequate supplies of goods is not the only difficulty to be overcome in order that a utopia may survive. Another requirement is that its citizens be willing to accept a common political philosophy, and this demands that their minds be molded according to the same pattern. Whereas this could not be achieved in the past, whatever the form of coercion, it is now brought about all over the world by subtle means of influencing the mind. Mass media are used in totalitarian states to shape political and social creeds, and in democratic countries they create a uniformity of habits and tastes. As tastes in material things cannot be dissociated in the long run from tastes in ideas and concepts, and therefore from beliefs and values, it may turn out that men will become so nearly uniform that they can readily be organized in utopian societies everywhere. The Brave New Worlds thus created will probably be as dull as the one visualized by Aldous Huxley but so, probably, would have been those imagined by Plato, Francis Bacon, Thomas More, William Morris, and H. G. Wells.

Granted an abundance of earthly goods and uniformity in political philosophy, there remains another stumbling block on the way to utopia, probably the most difficult of all to hurdle. It originates from the fact that nothing is stable in the world, neither natural resources nor the attitudes and tastes of men. The grapes that used to grow in England no longer ripen there. When water became scarce in the American Southwest and the trees in this area grew more slowly—as evidenced by the narrowness of the tree rings—crops failed and the Pueblo

Indians had to abandon many of their villages. Depending upon the distribution of rainfall, the Great Plains can be a farmer's haven or a dust bowl, the French vineyards yield *grands crus* or *vin ordinaire*.

Similarly, the responses of the human mind cannot help being modified by changes in environment; however well organized society may be, these changes are inevitable, and their effects are to a large extent unpredictable. Even the dogs most carefully prepared by Pavlov lost their conditioning when their environment was suddenly altered. Konrad Lorenz has shown that "imprinting," the behavior pattern acquired by birds early in life, can also be removed. The human body, too, changes with time, not only as a result of genetic alterations which may long remain unnoticed, but also because different societies cultivate different kinds of physical traits. The muscle development in the Athenian athlete was different from that deemed ideal today, and the obese woman regarded as a Venus at a certain time in certain parts of the world is repulsive to men in other eras and other climes.

The fundamental weakness of all ancient and of most modern utopias is that they postulate a more or less stable society in a stable environment; this is true of Plato's, More's, and Bacon's creations. But in fact static societies cannot survive because they are bound to crack and collapse in a world where everything else is endlessly changing and moving. Among utopists H. G. Wells was the first to realize that societies, like other living things, are in unstable dynamic equilibrium with their environment—an understanding he probably owed to his strong training in biological science under Thomas Huxley. In H. G. Wells's words,

The Utopia of a modern dreamer must needs differ in one fundamental aspect from the Nowheres and Utopias men planned before Darwin quickened the thought of the world. Those were all perfect and static States, a balance of happiness won forever against the forces of unrest and disorder that inhere in things. One beheld a healthy and simple generation enjoying the fruits of the earth in an atmosphere of virtue and happiness, to be followed by other virtuous, happy, and entirely similar generations, until the Gods grew weary. Change and development were dammed back by invincible dams forever. But the Modern Utopia must be not static but kinetic, must shape not as a permanent state but as a hopeful stage, leading to a long ascent of stages. Nowadays we do not resist and overcome the great stream of things, but rather float upon it. We build now not citadels, but ships of state.

Thus, it would seem that utopias can exist as realities only if they die shortly after being born, to be reborn with new shapes. Or, more probably, utopian dreams never come into being. They act as catalysts converting the crude materials of reality, the tools and products of experience and of science, into civilizations which take the shape developed first as an image in the mind of man. Utopias are like holy spirits which give the breath of life to matter.

The signers of the Declaration of Independence held it as a self-evident truth that the pursuit of happiness is an inalienable right of man. Although they did not define happiness, it can be taken for granted that, like most men, they believed a happy life implies the satisfaction of a few essential needs: enough of the right kind of food; adequate protection against the elements; freedom from disease; some variety of experience to keep the senses and the mind stimulated at the proper level; sufficient resources to avoid drudgery and to provide the objects of

H. G. Wells.

Low

This famous cartoon by Low represents H. G. Wells
at the height of his career—the extroverted,
brilliant, and learned exponent of science,
the fighting optimist of Men Like Gods and A Modern Utopia.
Less well known is the profound pessimism
of Wells's last published work.
In Mind at the End of Its Tether
he wrote that "Everything was driving
anyhow to anywhere at a steadily increasing velocity . . . .
The pattern of things to come faded away . . . .
The end of everything we call life is close at hand
and cannot be evaded."

one's desires; and above all, perhaps, the freedom to decide at every moment on the course of one's life.

It was such a blueprint for modern life that Condorcet had presented in his *"Esquisse d'un tableau historique des progrès de l'esprit humain,"* in which he traced the evolution of man from a barbaric stage to the age of reason. He outlined the future of man's progress on three lines: (1) the destruction of inequalities between nations; (2) the destruction of inequalities between classes; (3) the unlimited improvement of individuals mentally, morally, and physically. For the past two centuries the natural and social scientists of the Western World have labored to bring to reality these dreams of the philosophers of the Enlightenment.

Were they to come back to life, Condorcet and his contemporaries would probably be startled to find that most of the specific goals they visualized have now been reached, at least in the countries of Western civilization, and especially in the United States. Although we have become somewhat blasé about the marvels of our age, their magnitude can be recaptured by imagining how their effects on daily life would have excited the eighteenth-century mind. Medicine has come close to solving the problems of disease which as late as the nineteenth century made the average human life short and uncertain. Nutritional science has determined all the essential requirements of man, and technology has made it possible to satisfy them at all seasons in any climate. Everybody in the Western World can afford to keep warm during the winter and soon will be able to keep cool during the summer. Distances become every day less of a problem, and neither lack of time nor fear of physical strain need limit any longer our ability to move from one part of the globe to another.

At least as miraculous as the achievements of the natural sciences are those of politics, economics, and sociology. True enough, everybody bemoans the inefficiency of governments and the venality of politicians. But in fact most of the great social goals for which our ancestors were still fighting and dying a century ago have become the law of the land. We are in the process of becoming a classless society; in principle, if not always entirely in practice, we are free to express our opinions and to shape our destinies; Liberty and Equality are to a large extent *faits accomplis,* and even Fraternity may someday pass from our lips to our hearts.

The revolutionary advances of the past two centuries suggest that almost any problem of human welfare can be solved if it is properly formulated and if its solution is diligently pursued. As a student of experimental medicine, I take it for granted that progress can be made in the control of any disease to which we address ourselves with enough energy. I feel confident, also, that physicists, chemists, and engineers can provide us with almost any kind of earthly good. I even believe that sociologists and politicians will find ways of improving relations among men, even though the result may be peace without love. From penicillin to supersonic flight, from the control of personality to space exploration, from elimination of child labor to universal suffrage, the twentieth century has been marked by many scientific and social achievements which are so startling as to dwarf the miracles of the legendary ages.

Despite all these modern miracles, there are many

among us who speak regretfully of the old times and tend to place the golden age in the past rather than in the future. And, in fact, the many beautiful things that have come to us from the past are eloquent witnesses to a kind of happiness that we may well envy our ancestors: the lyrical outbursts of poets, the smiling angels in Gothic cathedrals, the glamorous feasts of the Renaissance, the gay celebrations of primitive country folk. How often we long for the profound and genuine happiness of yesteryear! But on the other side of the balance of time we have to place the sufferings that in the past remained unexpressed, the hunger that so often was the fate of poor people, the cold against which there was little protection, the pain that could not be relieved, the tortures that man inflicted on his fellow men, the sudden death of children and of adults, threatening alike the powerful of the earth and the unprivileged.

The disenchanted mood of today, of course, has its origin in the fact that happiness does not depend only upon comfort and contentment. Illiterates may well be contented and morons even more so—still more, perhaps, the proverbially contented cow or the well-fed household cat. But the further man evolves from his animal origin, the less happiness he can find in the mere removal of discomfort and in the satisfactions of the body. Every fulfillment, whatever its nature, is likely to create a new need and thus become a source of new dissatisfaction. The endless urge for some new experience, the tendency to look for goals beyond the attainable are traits which differentiate man from other forms of life. These aspirations have led him to establish his dominance over the natural world, but certainly they are endlessly creating for him new problems which make of health and of happiness mirages that are ever receding into the future. However, to the extent that happiness is conditioned by the elimination of physical suffering and the acquisition of certain forms of physical comfort, we have gone very far indeed toward fulfilling the dreams of utopists and reaching the goals formulated two centuries ago by the philosophers of the Enlightenment.

In his book *The Wonderful Century*, A. R. Wallace listed twenty-four revolutionary changes brought about by science during the nineteenth century as against fifteen for all preceding ages. If he were living today, Wallace might regard our time as the era of fulfillment. Ancient dreams of mankind which had remained utopian for millennia are now so nearly fulfilled that they are taken for granted by several hundred million people, and are claimed as a birthright by the rest of humanity. What adds to the wonder of this achievement is that it all happened so rapidly—in less than two hundred years—without any change in the innate abilities of man himself. The oral tradition, the written texts, the monuments that have come down to us from antiquity all bear witness to the fact that the human mind long ago acquired the ability to formulate highly abstract and sophisticated concepts, and to convert them into concrete realizations. The revolution in the ways of life occurred when man learned to apply the experimental method *systematically* to the material problems of everyday existence. As a result of this effort, technology began to grow geometrically some three hundred years ago, each invention and each improvement facilitating the next step. The acceleration of growth which was already apparent early in the nineteenth century is now breath-taking—at times, indeed, almost frightening. In the words of the

German physicist Max Born, "As I am almost eighty years old, I have myself experienced about half of this technical period. In my youth the difference in the way of life from that of, let us say, Caesar's time, appeared prodigious; but the differences between today's existence and that of my youth are incomparably greater."

In the countries of Western civilization where industrial technology was created, the increase of material wealth was accompanied by a progressively more equitable distribution of wealth, with the result that opportunities for rewarding individual effort were more widespread, social stratification was weakened, and political liberties were increased. Thus, it can be said that, indirectly but very effectively, the application of experimental science to technology has provided the concrete materials out of which utopian dreams were converted into reality.

Until the end of the eighteenth century most of material civilization had been built out of practices evolved either empirically from the very experience of day-to-day life or from discoveries made by accident without prior scientific knowledge. In fact, much of science itself arose from these empirical achievements. Then, systematic scientific knowledge derived from laboratory experimentation rapidly overtook practical life, and scientists increasingly became the innovators and indeed the governors of human existence. It can be said that the scientific age began when, from toiling obscurely in the rear of the empirical procedures, science stepped forward and held up the torch in front. By the middle of the nineteenth century, scientific investigation undertaken in a search for pure knowledge began to suggest practical applications and inventions. Faraday's electromagnetic experiments led to the dynamo and other electromagnetic machines; Maxwell's studies of waves led to wireless telegraphy; Pasteur's work revolutionized fermentation industries and the practice of medicine, etc., etc.

As a result of this change, the business of everyday life is now carried out with the tools provided by science, and, more importantly, the very character of human existence is now molded by the products of scientific technology. While these facts are obvious and acknowledged by all— by those who deprecate them as well as by those who delight in them—it is not so well recognized that the direction of scientific effort during the past three centuries, and therefore the whole trend of modern life, has been markedly conditioned by an attitude fostered by the creators of utopias. They fostered the view that nature must be studied not so much to be understood as to be mastered and exploited by man.

The urge to control nature is probably the most characteristic aspect of Western civilization. It has not yet been proven, however, that this ideal is the best for human life. After all, great civilizations have been created in the past, and much profound happiness has been experienced, based on the philosophy that man must strive for harmony with the rest of nature instead of behaving toward it as a dominating lord and an exploiting master. It is much too early to be sure that Galileo, Watt, and Edison have contributed more lastingly to human advancement and happiness than have Socrates, Lao-tze, and Francis of Assisi.

It is clear, of course, that the wealth and the conveniences which are the fruits of the Industrial Revolution are now accepted by all and enjoyed even by the enemies of science. But deep in the human conscience there is uneasiness about the enormous toll that mankind has paid for these advantages. There come to mind the misery and degradation of the labor classes in the industrial tenements of the nineteenth century, the excesses of child labor, the gross malnutrition and the destructive epidemics, the loss of the ancestral traditions which once added color and poetry to the life of everyone. Even today the ugliness of the industrial suburbs in the large cities of Europe and the destruction of natural beauty everywhere serve as reminders of the tragic truth that to achieve his purpose modern man has been willing to deface and rape nature.

It is true that, little by little, prosperous communities are trying to correct some of the blemishes left by two centuries of wanton exploitation and destruction. But lest we take too much pride in these halfhearted efforts, it is important to keep always in mind that today's prosperity was built out of the lives of countless millions of people and out of the desecration of our lands. Indeed, one may wonder whether man would have deliberately started the Industrial Revolution if he had been able to visualize beforehand what its costs and consequences would be. Anyone asking this question might well ponder the remarks on English scenery made by Jacquetta Hawkes in *A Land:*

Recalling in tranquillity the slow possession of Britain by its people, I cannot resist the conclusion that the relationship reached its greatest intimacy, its most sensitive pitch, about two hundred years ago. By the middle of the eighteenth century, men had triumphed, the land was theirs, but had not yet been subjected and outraged. Wildness had been pushed back to the mountains, where now for the first time it could safely be admired. Communications were good enough to bind the country in a unity lacking since it was a Roman province, but were not yet so easy as to have destroyed locality and the natural freedom of the individual that remoteness freely gives. Rich men and poor men knew how to use the stuff of their countryside to raise comely buildings and to group them with instinctive grace. Town and country having grown up together to serve one another's needs now enjoyed a moment of balance.

The first Industrial Revolution is now a *fait accompli*, and it is idle to ask whether men would have refrained from starting it, or would have directed its course more wisely, had they been able to foresee the distant consequences of their acts. But it is urgent that we have these questions in mind as we enter the era of the second Industrial Revolution. One of the reassuring aspects of the modern world is that much thought is now being given to the possible effects of the future applications of science. For example, exhaustive studies of the biological effects of ionizing radiation were undertaken before atomic energy became an industrial source of power. The obvious threats of atomic warfare have, of course, been widely publicized, but of greater interest are the current discussions centering on the possible dangers of the application of atomic energy to peaceful purposes—for example, with regard to the contamination of water supplies.

The alarming increase in world population is another problem which is being considered from many different angles. The development of universally acceptable techniques for birth control and of policies for increasing food production are naturally the aspects of the problems which first come to mind and are the most widely discussed. But other considerations at least as important are being brought forward as a result of more searching analysis. Thus, it is becoming clear that one must be concerned not only with the number of human beings on earth, but also with the quality of their bodies and their minds. As is discussed in the following chapter, birth control alone might lead to biological disaster if not guided by genetic considerations and supplemented by the proper physiological training of people at all ages.

It is becoming apparent also that the increase in population has consequences which transcend the supplies of food and of raw materials. There is no doubt that, by taking full advantage of the potentialities of agricultural and industrial technologies, it would be possible to take care of a world population much larger than that which now exists. But if man continues to proliferate, he will have to eliminate from the globe the other living things that compete with him for food and space. Continued increase in human population spells the virtual extinction of all wildlife—a dreary prospect indeed! As people become more numerous, furthermore, there is bound to be increasing reglementation amounting to regimentation, a spread of uniformity, and progressive disappearance of most individual liberties. Foods and manufactured commodities are not the only factors to be considered in a measurement of the standard of living. Beauty, independence, solitude also have always been among the essential requirements for civilized life.

Many other examples could be adduced to illustrate that problems of crucial importance for mankind which appear at first sight predominantly technical involve in reality judgments of values. Science itself, in its pure form, is concerned with phenomena and laws on which all qualified and thoughtful persons eventually agree irrespective of their individual beliefs, opinions, and tastes.

But, unlike pure knowledge, the applications of science are concerned with the desires of mankind, its whims and fancies, as much as with its biological needs. As stated by J. M. Clark, "There are two worlds, the world of impersonal investigation of cause and effect, and the world of desires, ideals and value judgments. The natural sciences deal with the first, ethics with the second." The physicist who works on the industrial applications of any form of energy cannot help becoming involved in problems of ethics; and this is true also of the physician and the biologist working on medical problems—as we shall see in the following chapter.

The ethical problems posed by the utilization of knowledge are, of course, as old as mankind. But it is only during modern times that the question has become practically important as a result of the increasing effectiveness of scientific methods, and of the fact that science is now valued more for its social uses than as natural philosophy. In a recent address Ritchie Calder stated that "Scientists leave their discoveries like foundlings on the doorstep of society. The step-parents do not know now to bring them up." Clearly, this attitude is no longer permissible now that scientific discoveries can have such far-reaching and lasting effects on human existence—indeed, on the fate of the human race. It is for society, of course, to decide what goals it wishes to reach and what risks it is willing to take. But it is the task of the scientific community to formulate as clearly as possible and to make public the probable consequences of any step that it takes and of any action that it advocates. In other words, the responsibility of the scientist does not stop when he has developed the knowledge and techniques that lead to a process or a product. Beyond that, he must secure and make public the kind of information on which the social body as a whole can base the value judgments that alone will decide long-range policies.

The relation of science to society has changed and become more complex during modern times. Three hundred years ago Bacon and his followers were justified in claiming that the important problem was to learn *how* to do things. There was then so little that could be done! Soon it became apparent that the most effective method of progress was to try to understand natural phenomena, their *whys* as much as their *hows*. Now it can be said that it is possible to achieve almost anything we want—so great is the effectiveness of technology based on the experimental method. Thus, the main issue for scientists and for society as a whole is now to decide *what* to do among all the things that could be done and should be done. Unless scientists are willing to give hard thought —indeed, their hearts—to this latter aspect of their social responsibilities, they may find themselves someday in the position of the Sorcerer's Apprentice, unable to control the forces they have unleashed. And they may have to confess, like Captain Ahab in *Moby Dick,* that all their methods are sane, their goal mad.

4

Medical Utopias

In medicine even more than in other fields of science, theories and practices have always been under the sway of *a priori* philosophical attitudes and rationalized beliefs. The social forces that have influenced medical history range from the primitive fear of demons to the current wave of faith healing, from Rousseau's assertion that "hygiene is less a science than a virtue" to the modern illusion that diseases can be conquered by drugs.

Among all the medical utopias that have flourished in the course of time, none has blossomed so constantly and in so many forms as the belief that disease can be entirely eliminated from the earth. At the present time this illusion is based on an uncritical faith in the magic power of experimental science. But fundamentally it arises from the mystical belief in the existence of what Thomas De Quincey, in "Confessions of an English Opium-Eater," called "that sort of vital warmth...

which would probably always accompany a bodily constitution of primeval and antediluvian health."

Many thinkers of classical Greece certainly believed that reasonable men could achieve the millennium of health by the exercise of wisdom. Witness the cult of Hygeia, which was an expression of the faith that men could enjoy *mens sana in corpore sano* if they lived according to reason. Carrying this doctrine to its logical conclusion, Plato wrote that the need for many hospitals and doctors was the mark of a bad city; there would be little use for them in his ideal Republic. In Imperial Rome, Tiberius asserted in a similar vein that anyone who consulted a doctor after the age of thirty was a fool for not having yet learned to regulate his life properly without outside help. Although medieval Christianity had little faith in the possibility of creating a paradise of health on earth, Thomas More and all the utopists that followed him after the Renaissance popularized imaginary states so well organized that their medical needs could be foreseen and provided for just as certainly as their political and economic problems. Describing the ideal society he imagined on the moon, Cyrano de Bergerac asserted, "In every house there is a Physionome supported by the state, who is approximately what would be called among you a doctor, except that he only treats healthy people."

The French Encyclopedists believed that all health problems could be solved by science, and Condorcet envisaged a rational world free from disease, in which old age and death would be infinitely postponed. Echoing this faith, Benjamin Franklin wrote to the English chemist Joseph Priestley that "all diseases may by sure means be prevented or cured, not excepting that of old age, and our lives lengthened at pleasure even beyond the antediluvian standard." Fourier was more specific, asserting that in the society of his dreams man's life would be prolonged to 144 years. Continuing the traditions of the Enlightenment, Rudolf Virchow preached in his journal *Medizinische Reform* that misery was the breeder of disease, and that the key to the general improvement of health would be found in the improvement of social conditions. In one form or another, projections of utopia have continued until our time. In James Hilton's *Lost Horizon* the lamas living in Shangri-La, miles from corrupting influences, had mastered the secret of long life. In his book *My First Years in the White House* Huey Long listed high on his program a plan to provide adequate medical care for the whole country—giving the job to the Mayo brothers! And throughout the 1960 presidential campaign, protection from womb to tomb seemed to overshadow most other political issues.

Faith in the powers of man to eradicate disease has been greatly strengthened, of course, by the spectacular scientific achievements of the nineteenth century. Early in the twentieth century Hermann Biggs, then Commissioner of Health of New York State, adopted for his department the motto "Public Health Is Purchasable. Within Natural Limitations Any Community Can Determine Its Own Death Rate." In 1958 the same faith was repeatedly expressed on the occasion of the tenth anniversary of the World Health Organization. The authors of the WHO pamphlet *Ten Years of Health Progress* recognized that large problems remained to be solved, and that "as one disease is eradicated . . . others grow in importance," but Dr. Axel Hojer voiced their collective confidence that through techniques based on scientific knowledge "man seems to have found out how to make his dreams of a paradise on earth come true." Medical scientists may be skeptical about social utopias designed on the basis of political theories, but they rarely doubt that mankind would soon achieve the millennium if their own theories based on natural sciences were put into practice!

The widespread conviction that health is purchasable, not only in limited areas but also on a world-wide scale, seems to be justified by the advances made during the past half-century in the fields of nutrition and infection. In reality, however, it has not yet been shown that these achievements justify the wide extrapolations made from them. There is overwhelming historical evidence that the evolution of diseases is influenced by many determining factors which are not as yet amenable to social or medical control, and may never be. The changes that have occurred without benefit of conscious human intervention in the prevalence of various diseases during the past few centuries should serve as a warning that it is unwise to predict the future from the short perspective of the past decades.

Granted the lack of precise information, it is clear that there have been spontaneous ebbs and flows in the prevalence and severity of many diseases. Plague invaded the Roman world during the Justinian era; leprosy was prevalent in western Europe until the sixteenth century; plague again reached catastrophic proportions during the Renaissance; several outbreaks of the sweating sickness terrorized England during Tudor times; syphilis spread like wildfire shortly after 1500; smallpox was the scourge of the seventeenth and eighteenth centuries; tuberculosis, scarlet fever, diphtheria, measles took over when smallpox began to recede; today virus infections occupy the focus of attention in our medical communities; and long before viruses had become scientifically fashionable, pandemics of influenza at times added a note of still greater unpredictability to the pattern of infection.

Awareness that diseases come and go for mysterious reasons is not new. Malthus had sensed the phenomenon when he wrote in 1803, "For my part, I feel not the slightest doubt that, if the introduction of the cow pox should exterminate the small pox, we shall find a very perceptible difference in the increased mortality of some other disease." More recently the historical and geographic aspects of the problem were documented by August Hirsch in his monumental *Handbook of Geographical and Historical Pathology*. The matter was interestingly discussed by Charles Anglada in *Études sur les maladies éteintes et sur les maladies nouvelles*, and by Charles Nicolle in his famous book *Naissance, vie, et mort des maladies infectieuses*, where he showed that the biological aspects of life are governed by forces independent of conscious human intervention. Most explicit perhaps was the statement made in 1873 by William Parr in his letter to the Registrar General in England:

The infectious diseases replace each other, and when one is rooted out it is apt to be replaced by others which ravage the human race indifferently whenever the conditions of health are wanting. They have this property in common with weeds and other forms of life, as one species recedes another advances.

I have selected infection to illustrate ebbs and flows in the prevalence of disease because of my greater familiarity with this field. But anyone with specialized knowledge could provide just as telling examples from other areas of medicine. With regard to nutrition, Lucretius was already aware of the problem when he wrote two thou-

The style and subject matter of this cartoon
place it in the middle of the nineteenth century in England.
Gout, tisis (tuberculosis), and colic (intestinal disorders)
were then very common in all social classes.
It is interesting to note that, while these diseases still occur,
they are now much less prevalent in the Western World.
This change is due not so much to therapeutic advances
as to modifications in the ways of life.

sand years ago, "In the old days lack of food gave languishing limbs to Lethe. On the contrary, surfeit of things stifles us today."

Coming now to our times, who could have dreamed a generation ago that hypervitaminoses would become a common form of nutritional disease in the Western World; that the cigarette industry, air pollutants, and the use of radiations would be held responsible for the increase in certain types of cancer; that the introduction of detergents and various synthetics would increase the incidence of allergies; that advances in chemotherapy and other therapeutic procedures would create a new staphylococcus pathology; that alcoholics and patients with various forms of iatrogenic diseases would occupy such a large number of beds in the modern hospital?

Despite the dark spots on the health picture of the modern world, many facts seem to provide support for those who claim that we are approaching medical utopia. Franklin's optimistic prophecy that our lives can be lengthened at pleasure even beyond the antediluvian standard appears close to fulfillment. Not so long ago most children died during infancy or in their teens; only a small percentage survived into adulthood. For example (according to Hugh Smith in *The Family Physician*), approximately two thirds of the children born in London between 1762 and 1771 died before the age of five, and fully three quarters of these never reached the age of two. That this state of affairs prevailed not only among the socially unfortunate is illustrated by the case of Queen Anne, with seventeen pregnancies and not a single surviving child.

The situation is very different today in the Western

World. Largely because of the control of microbial diseases and the improvements in general nutrition, the immense majority of children survive. Moreover, even those suffering from diabetes, congenital heart abnormalities, and many other formerly fatal diseases can now have an almost normal span of life. True enough, progress has been much less dramatic with regard to the diseases which affect adulthood. Yet even in this age group several types of ailments and accidents which once were uniformly fatal can now be treated so effectively that they have been all but eliminated as causes of death. As a result, more and more persons in our communities exceed the life span of threescore and ten to which so few could aspire in the past. While the countries of Western civilization have, of course, been the chief beneficiaries of the modern advances in human health, the less privileged parts of the world have shared in the fruits of medical science, and this fact has contributed to the increase in population all over the world.

The achievements of modern medical science are, indeed, almost miraculous. Surgery restores to function broken limbs and damaged hearts with amazing safety and little suffering; sanitation removes from our environment many of the germs of disease; new drugs are constantly being developed to relieve physical pain, to help us sleep if we are restless, to keep us awake if we feel sleepy, and to make us oblivious of worries. There are many good reasons, therefore, to share with Dr. M. G. Candau, Director General of WHO, the belief that the world will soon be in a position to make a reality of medical utopia. In his words,

If the great advances gained in science and technology are put at the service of all people of the world, our children and their children will live in an age from which most of the diseases our grandparents and parents took for granted will be banished. . . . It may no longer be Utopian to envisage a new chapter in the history of medicine.

No one can take exception to these words except to point out that they do not tell the whole story. Within limits it is true that we can control many of the diseases that "our grandparents and parents took for granted." But this does not prove that we know how to control the disease problems that will be encountered in the future, or even, for that matter, those of the present. It is seldom recognized that each type of society has diseases peculiar to itself—indeed, that each civilization creates its own diseases. Furthermore, there is no evidence that the techniques developed for dealing with the disease problems of one generation can cope with the problems of another. The word "Utopian" used by Dr. Candau is here the telltale. Whether they be medical or political, utopias imply, as pointed out earlier, a static view of the world. In reality, however, societies are never static. Nothing is stable in the world—men change, and so do all their problems. We can indeed expect a "new chapter in the history of medicine," but the chapter is likely to be as full of diseases as its predecessors; the diseases will only be different from those of the past.

There is no doubt, of course, that scientific medicine is now essential to our social existence and contributes much to the success of modern societies. The paradox is that, despite the spectacular advances in the knowledge and treatment of disease, the need for hospital facilities and the cost of medical care continue to increase. This results in part from the much more exacting criteria of health that prevail in modern societies, but other reasons are also apparent. While a few of the old disease problems are being solved, new ones are constantly cropping up and make modern man increasingly dependent on medicine for his very survival. The situation is not unlike that faced by the farmers and gardeners in their attempts to protect crops against weeds and pests. The task is never finished because as one problem is solved, another soon appears which requires attention. Nature always strikes back. It takes all the running we can do to remain in the same place.

Since no one can predict the disease patterns of the future, it will suffice to contrast here, for the sake of illustration, some of the problems which dominated medicine a century ago with those likely to increase in importance in the near future.

Disease from contaminated food or beverages was very common a few generations ago, and nutritional deficiencies were almost the rule. Now laboratories check on the safety of what we eat and drink. Furthermore, the nutritional requirements of man are now well known, and in the Western World, at least, we have the means to satisfy them. But all this theoretical and practical knowledge does not guarantee that nutrition will not present problems in the immediate future, even assuming that economic prosperity continues. On the one hand, modern agriculture and food technology have come to depend more and more on the use of chemicals to control pests and to improve the yields of animal and plant products. The cost of food production would enormously increase without these chemicals, and for this reason their use is justified. Unfortunately, however, and despite all care, several of them eventually reach the human consumer in objectionable concentrations. As more and more substances are introduced in agriculture and food technology every year, it will become practically impossible to test them all with regard to long-range effects on human health, and the possibility of toxic reactions must be accepted as one of the inevitable risks of progress.

Another potentially disturbing aspect of nutrition in the future arises from the fact that human food requirements have been formulated on the basis of a certain level of physical activity and of exposure to the inclemencies of the weather. But no one knows exactly what these requirements will be for the wheel-borne, air-conditioned human being of the next decades. The well-fed child of today may prove to have been overfed in view of the kind of life he will lead tomorrow.

Just as we have eliminated from our life physical overexertion and minimized exposure to heat or cold, so we have greatly decreased contact with infectious agents and learned to treat the diseases caused by some of those that we cannot yet escape. One needs only to recall the toll exacted by infectious diseases in the past to appreciate the magnitude of the contributions made to human health by sanitation and drug therapy. But even this achievement may turn out to be a not unmixed blessing. True enough, certain acute infectious diseases are practically under control, but others are becoming more prevalent—paralytic poliomyelitis, for example. And, more than ever before, hospitals are crowded with patients suffering from chronic disorders which do not kill outright but often ruin life.

Fifty years ago horses were essential for short-range transportation, and we had stables in every block. The stables constituted a fine breeding ground for flies, and these—as well as other insects—acted as vectors in the transmission of many diseases. Progressively mankind managed to cope with this problem by developing certain kinds of natural immunity and by introducing sanitary practices. Today the situation is different. Horses have been displaced by automobiles, and the diseases they brought about have therefore disappeared. But unsought results of this advance have been the pollution of the atmosphere by automobile exhausts, and forty thousand fatalities each year on our highways. We are beginning to think constructively about these new problems and may learn to do something about them, but we may expect that new difficulties will soon arise from further improvements in methods of transportation. Probably unexpected toxic manifestations will accompany the use of new fuels or of atomic energy. Probably also a few more of our degrees of freedom will be eliminated by the enforcement of more drastic traffic laws, and this will contribute still further to the soul-destroying mechanization of our everyday life.

In any case, the rapid increase in population—the population explosion, as it is properly called—will make it more and more difficult to avoid regimentation of life. Even though political freedom may survive in principle, individual activities will be increasingly restricted within any political system. The social structure is bound to become more rigid as the larger population becomes increasingly dependent on a more complex technology. Until we have learned to change our ways, restrictions of individual freedom are likely to increase the incidence of psychoneuroses.

In the same class belong the mental health problems caused by automation. While some look upon automation as the gateway to the golden age, others see it as the source of new social and medical difficulties. It is likely that automated work will induce in the worker a sense of inadequacy because of a feeling that he is less useful in an effort the significance of which he cannot assess; to anxiety arising from the more abstract character of his effort; and to strains resulting from new kinds of responsibility. Nervous strain will inevitably result from jobs that involve little physical activity but demand unremitting attention to signs of varying perceptibility, often separated by long intervals. There is no hope, in my opinion, that pills can prevent or cure these minor but important mental illnesses, or that medical personnel and services will ever be sufficient to deal with the huge numbers of persons who will need psychiatric help in the regimented life that will characterize the first phases of the automation age. Adjustments will certainly be made, but we can anticipate a period when boredom and mental frustration will have consequences worse than the scars that used to result from inclement weather or physical exertion.

It would be presumptuous on my part, and indeed impossible for anyone, to predict what precise effects the ever changing social environment will have on the children who are growing up today. But the formulation of the problem might be facilitated by considering one particular aspect on which some concrete information is available. As anyone can observe, most children are now growing in size more rapidly than did those of one or two generations ago. This change affects not only size but other physiological characteristics as well; in general, boys and girls are maturing physically and reaching puberty somewhere between six months and two years earlier than did their parents. It is probable that improvements in nutrition, control of infection, and other unidentified factors are responsible for this physiological acceleration, but of greater importance than the causes of acceleration are the problems that it poses for the management of young people.

It is ironic indeed that legislation and social mores tend to prolong the period during which young people are treated as if they were immature and irresponsible, precisely at the time when children are growing faster and physically maturing earlier. Vigorous and well-fed young people need rough physical activity, while society urges on them a sheltered and effortless life. They are eager to show their worth and to function usefully, while labor laws bar them from employment. They crave an imaginative life and the chance to manifest initiative, while most forms of responsibility are denied them because they are regarded as children.

It would be entertaining, if it were not tragic, to contrast the place occupied in our society by the modern fully developed six-foot teen-agers with that occupied by their physiological equivalents in the past. Throughout history young adults have acted effectively as leaders in warfare, active members of political parties, creators of business enterprises, or advocates of new philosophical doctrines—whereas modern young people are expected to find fulfillment in playgrounds, juvenile spectacles, and ice cream parlors! I am afraid that the present social attitude renders them more likely to yield to sexual impulses, become juvenile delinquents, or turn to philosophies of despair. New health problems are bound to occur among teen-agers if society does not recognize that fully developed, well-fed young bodies need satisfying and worth-while forms of expression in order to remain creative and healthy.

Most unexpected, perhaps, is the fact that medicine is creating new disease problems by reason of its very successes. This side of the picture came out in the course of a debate held in London in November, 1952, when the Hunterian Society voted 59 to 47 "that the continued advance in medicine will produce more problems than it solves." Facetious as it was, the resolution clearly reflected the awareness that all is not as well as advertised in the world of medicine.

Among the new problems arising from the partial control of man's ancient plagues, some have an economic basis and are apparent chiefly among the underprivileged peoples of the world. In the past, microbial diseases acted as one of the checks in population size by killing large numbers of children and keeping down the number of persons who reached old age. To put it crudely, nature balanced the books for us. Today, public health measures, supplemented by the use of insecticides and antimicrobial drugs, greatly reduce the number of early deaths. As the birth rates show as yet no sign of decreasing, the world population is growing in an unprecedented manner. In Ceylon, for example, the partial control of malaria resulting from the use of DDT has resulted in a sudden increase of the population during the past few years.

Unfortunately, the results of infection control are not an unmixed blessing, especially as some national economies are not capable of making orderly adjustments to the new state of affairs. It is of particular importance in this regard that food production and especially the supply of well-balanced proteins cannot keep pace with the increase in world population. Advances in agricultural and industrial technology will, of course, improve food supplies, but not fast enough for the needs. While the world population increased by eight per cent during the past ten years, food production increased by only five per cent, and in consequence more people go hungry today than did a decade ago. It is to be feared that food deficiencies will in the long run cause more physiological misery and suffering than have been prevented by the partial control of infection.

Less apparent as yet, but also constituting a threat for the future, are the economic and biological consequences of the survival of persons suffering from various types of physical and mental deficiency. The availability of techniques capable of postponing death in every age group and arresting almost any type of disease will increasingly present to the medical conscience difficult alternatives. For example, to save the life of a child suffering from some hereditary defect is a humane act and the source of professional gratification, but the long-range consequences of this achievement mean magnified medical problems for the following generations. Likewise, prolonging the life of an aged and ailing person must be weighed against the consequences for the individual himself and also for the community of which he is a part. These ethical difficulties are not new, of course, but in the past they rarely presented issues to the medical conscience because the physician's power of action was so limited. Soon, however, ethical difficulties are bound to become larger as the physician becomes better able to prolong biological life in individuals who cannot derive either profit or pleasure from existence, and whose survival creates painful burdens for the community. Increasing numbers of these persons cannot pull their full weight in society and require constant medical supervision and economic assistance. They constitute a social burden which is likely to grow heavier with time, precisely as a result of medical advances.

Even more important than these economic considerations, however, is the fact that many of the biologically defective individuals who are saved from death transfer to their progeny the genetic basis of their deficiencies. Speaking of our "load of mutations," Professor H. J. Muller has repeatedly emphasized that, as medical science becomes more effective in prolonging survival, there will be an increase in the frequency of detrimental genes allowed to accumulate in our communities. A continua tion of this trend would mean that, in Professor H. J. Muller's words,

Instead of people's time and energy being mainly spent in the struggle with external enemies of a primitive kind such as famine, climatic difficulties, and wild beasts, they would be devoted chiefly to the effort to live carefully, to spare and to prop up their own feeblenesses, to soothe their inner disharmonies and, in general, to doctor themselves as effectively as possible. For everyone would be an invalid, with his own special familial twists.

In a recent essay on "The Control of Evolution in Man," the English geneticist C. D. Darlington expressed tersely the same thought:

Those who were saved as children return to the same hospital with their children to be saved. In consequence, each generation of a stable society will become more dependent on medical treatment for its ability to survive and reproduce.

Let us hasten to say that not all geneticists take such a dark view of our biological future in the Western World.

But even the most optimistic probably recognize that there is much truth in a quatrain published shortly after World War I in the London *Spectator:*

> Science finds out ingenious ways to kill
> Strong men, and keep alive the weak and ill—
> That these a sickly progeny may breed,
> Too poor to tax, too numerous to feed.

While these problems are becoming more apparent in modern societies, they are not entirely new. In ancient civilizations and among primitive peoples one commonly finds social customs and taboos which represent efforts to deal with the difficulties created by the existence of individuals who are economically or biologically deficient. Among Eskimos and certain Indian tribes aged people were expected to abandon the camp and die once they had become too much of a burden for the group. For different but related reasons, unwanted children in Sparta were left to die from exposure.

Similar practices which seem cruel to us have been defended by some of the most idealistic social philosophers in the past. Suicide of the sick was encouraged by the Stoics and other sects, and even by Plato. At Marseilles in Roman times poison was kept in the city for those who could present to the Council a reason for wishing to rid themselves of life. More surprisingly, a similar attitude was taken in the seventeenth century by St. Thomas More. In More's Utopia those who were incurably ill and in continual pain were urged by the priests and magistrates to kill themselves.

Yf the dysease be not onelye vncurable, but also full of contynuall payne and anguyshe, then the priestes and the magistrates exhort the man, seynge he ys not able to doo annye dewtye of lyffe, and by ouerlyuing hys owne deaths is noysome and yrkesome to other, and greuous to hymself; that he wyll determyne with hymselfe no longer to cheryshe that pestilent and peynefull dysease: and, seynge hys lyfe ys to hym but a tourmente, that he wyll nott be vnwyllynge too dye.

In this respect at least, the ethics of the modern world have grown loftier since Plato and More. All decent men now regard human life as sacred, worth preserving at whatever cost to the individual or to society. Moreover, experience has taught us that some of the greatest achievements of the human race have come from individuals suffering from handicaps which rendered their physical lives miserable and would have prevented their survival in an unsheltered environment. Thus, selfish motives agree with modern ethics and religious ideals in making the preservation and prolongation of human life the ultimate goal of medicine.

There is, furthermore, biological jusification for this attitude, for it is misleading to speak of biological defectives without regard to the environment in which these individuals live and function. Medical techniques can make up for genetic and other deficiencies that would be lethal in the wilderness. While it is certain that the physically handicapped could not long survive under "natural" conditions, it is also true that medical and other social skills make it possible for men to live long and function effectively in the modern world even though they be tuberculous, diabetic, blind, crippled, or psychopathic. The reason is that fitness is not an absolute characteristic, but rather must be defined in terms of the total environment in which the individual has to spend his life.

It must be realized, however, that fitness achieved through constant medical care has grave social and economic implications which are commonly overlooked. We can expect that the cost of medical care will continue to soar because each new discovery calls into use more specialized skills and expensive items. Today medical care represents some eight per cent of the national income in the United States. There is certainly a limit to the percentage of society's resources that can be devoted to the maintenance of medical establishments, and a time may come when medical ethics and policies will have to be reconsidered in the harsh light of economics.

Furthermore, it must not be taken for granted that the power of science is limitless. Only during the past few decades and in but a few situations has medical treatment enabled the victims of genetic disabilities to survive and to reproduce on a large scale. If the numbers of biologically defective individuals continues to increase, therapy may not be able to keep pace with the new problems that will inevitably arise. Yet, failure to meet these problems might eventually lead to biological extinction. It is urgent, therefore, that we formulate medical policies compatible with our system of ethics, yet practical within the limitations imposed by economic and biological consequences.

For the reasons outlined above, and despite all we can do in the way of medical care, new problems of disease will endlessly arise and require ever increasing scientific and social efforts, making of medical Utopia a castle in the air that can exist only in the Erewhon of political Utopia. Yet, it is clear that the lay and paramedical organizations established during the past fifty years to deal with problems of health are based on the optimistic assumption that, given enough time and financial resources, science can develop techniques to prevent or cure most diseases, and that only social and economic limitations will in the future stand in the way of ideal health. Anyone who has dealt with congressional appropriation committees knows that their willingness to pour public money into medical research comes from the belief that if the program is pursued for a few more years, science will provide ways to eliminate disease.

The belief that disease can be conquered through the use of drugs deserves special mention here because it is so widely held. Its fallacy is that it fails to take into account the difficulties arising from the ecological complexity of human problems. Blind faith in drugs is an attitude comparable to the naïve cowboy philosophy that permeates the Wild West thriller. In the crime-ridden frontier town the hero single-handedly blasts out the desperadoes who have been running rampant through the settlement. The story ends on a happy note because it appears that peace has been restored. But in reality the death of the villains does not solve the fundamental problem, for the rotten social conditions which opened the town to the desperadoes will soon allow others to come in unless something is done to correct the primary source of trouble. The hero moves out of town without doing anything to solve this far more complex problem; in fact, he has no weapon to deal with it and is not even aware of its existence.

Similarly, the accounts of miraculous cures rarely make clear that arresting an acute episode does not solve the problem of disease in the social body—or even in the individual concerned. Gonorrhea in human beings has been readily amenable to drug therapy ever since 1935; its microbial agent, the gonococcus, is so vulnerable to penicillin and other drugs that the overt forms of the disease can now be arrested in a very short time, and at a very low cost. Yet gonorrhea has not been wiped out in any country or social group. The reason is that its control involves many factors, physiological and social, which are not amenable to drug treatment. These factors range all the way from the ill-defined conditions which allow the persistence of gonococci without manifestation of disease in the vagina of "successfully" treated women, to the economic and psychological aspects of the social environment which favor loose sexual mores and juvenile delinquency.

To state it bluntly once more, my personal view is that the burden of disease is not likely to decrease in the future, whatever the progress of medical research and whatever the skill of social organizations in applying new discoveries. While methods of control can and will be found for almost any given pathological state, we can take it for granted that disease will change its manifestations according to social circumstances. Threats to health are inescapable accompaniments of life.

Health is an expression of ability to cope with the various factors of the total environment, and fitness is achieved through countless genotypic and phenotypic adaptations to these factors. Any change in the environment demands new adaptive reactions, and disease is the consequence of inadequacies in these adaptive responses. The more rapid and profound the environmental changes, the larger the number of individuals who cannot adapt to them rapidly enough to maintain an adequate state of fitness and who therefore develop some type of organic or psychotic disease. "It is changes that are chiefly responsible for diseases," wrote Hippocrates in *Humours*, "especially the great changes, the violent alterations both in seasons and in other things." And he stated again in *Regimen in Acute Disease,* "The chief causes of disease are the most violent changes in what concerns our constitutions and habits."

A perfect policy of public health could be conceived for colonies of social ants or bees, whose habits have become stabilized by instincts. Likewise, it would be possible to devise for a herd of cows an ideal system of husbandry with the proper combination of stables and pastures. But unless men become robots, their behavior and environment fully controllable and predictable, no formula can ever give them permanently the health and happiness symbolized by the contended cow. Free men will develop new urges, and these will give rise to new habits and new problems, which will require ever new solutions. New environmental factors are introduced by technological innovations, by the constant flux of tastes, habits, and mores, and by the profound disturbances that culture and ethics create in the normal play of biological processes. It is because of this instability of the physical and social environment that the pattern of disease changes with each phase of civilization, and that medical research and medical services cannot be self-limiting. Science provides methods of control for the problems inherited from past generations, but it cannot prepare solutions for the specific problems of tomorrow because it does not know what these problems will be. Physicians and public health officials, like soldiers, are always equipped to fight the last war.

What may be worth asking is whether medical science can help the individual and society to develop a greater ability to meet successfully the unpredictable problems of tomorrow. This is an ill-defined task for which there is hardly any background of knowledge. Traditionally, medicine is concerned with retarding death and also with preventing pain and minimizing effort. Its achievements in this field have added greatly to the duration, safety, and charm of individual existence. While scientific medicine has continued to emphasize the detailed study of particular diseases and specific remedies, it has placed less emphasis on the nonspecific mechanisms by which the body and soul deal with the constant and multifarious threats to survival. The question is whether it is possible to increase the ability of the individual and of the social body to meet the stresses and strains of adversity. In this regard it may be worth considering that preoccupation with the avoidance of threats and dangers does not have the creative quality of goal-seeking. It is at best a negative attitude, one that does not contribute to growth, physical or mental. In our obsession with comfort and security we have given little heed to the future, and this negligence may be fatal to society and, indeed, to the race.

Whatever the theories of physicians, laboratory scientists, and sociologists, it is of course society that must decide on the types of threats it is most anxious to avoid and on the kind of health it wants—whether it prizes security more than adventure, whether it is willing to jeopardize the future for the sake of present-day comfort. But the decision might be and should be influenced by knowledge derived from a study of the manner in which different ways of life can affect the future of the individual and of society. Although this knowledge does not yet exist, a few general remarks appear justified.

It is a matter of common experience that, while man's physical and mental resources cannot develop to the full under conditions of extreme adversity, nevertheless a certain amount of stress, strain, and risk seems essential to the full development of the individual. Normal healthy human beings have long known, and physiologists are beginning to rediscover, that too low a level of sensory stimulation may lead to psychotic disorders, and that man functions best when a sufficient number of his neurons are active. Analogous considerations seem to be valid for the lower levels of biological functions, and recent studies illustrate that at least some of the mechanisms involved in training and in adaptability are not beyond experimental analysis.

It has been shown by Dr. Curt P. Richter and his associates that the domesticated laboratory rat differs from its wild ancestor, the Norway rat, in many anatomic and physiologic characteristics that can be measured by objective tests. As a result of selection and of life in the sheltered environment of the laboratory, the domesticated rat has lost most of its wild ancestor's ability to provide for itself, to fight, and to resist fatigue as well as toxic substances and microbial diseases. The domesticated rat has become less aggressive in behavior but also less able to meet successfully the strains and stresses of life, and therefore it could hardly survive competition in the free state. As a result of domestication, in Dr. Richter's words:

(1) the adrenal glands, the organs most involved in reactions to stress and fatigue, and in providing protection from a number of diseases, have become smaller, less effective . . . (2) the thyroid—the organ that helps to regulate metabolism, has become less active . . . (3) the gonads, the organs responsible for sex activity and fertility, develop earlier, function with greater regularity, bring about a much greater fertility . . . The finding of a smaller weight of the brain and a greater susceptibility to audiogenic and other types of fits, would indicate that the brain likewise has become less effective.

It must be pointed out, on the other hand, that the domesticated rat is better adapted than its wild ancestor to laboratory life and to many of the artificial tricks and stresses devised by the experimenter.

While some of these changes may be phenotypic, it is probable that most of them are the expression of mutations selected by life in the laboratory. But, whatever their mechanism, the effects of domestication on the wild rat are not without relevance to the future of mankind. Human societies made up of well-domesticated citizens, comfort-loving and submissive, may not be the ones most likely to survive.

The study of so-called germ-free animals has revealed other aspects of this problem. Animals born and raised in an environment free of detectable microorganisms can grow to a normal size and are capable of reproducing themselves for several generations, but they exhibit a high susceptibility to infection, even to the most common types of microorganisms that would be innocuous for animals raised in a normal, exposed environment. Furthermore, germ-free animals produce only small amounts of lymphoid tissue, and their plasma is extremely low in gamma globulin—deficiencies which may be of little consequence in the protected environment of the germ-free chamber, but which are great handicaps under normal conditions of life.

Experimental situations of these types illustrate the fact that a sheltered life alters in many ways the ability of the organism to cope with the stresses of life. "Let a man either avoid the occasion altogether, or put himself often to it, that he may be little moved with it," Bacon wrote in his essay "Of Nature in Men." While Bacon's aphorism is a picturesque statement of an important sociomedical problem, the solution that it offers hardly fits the modern world. Man cannot "put himself often to threats" whose nature he cannot anticipate. But he can perhaps cultivate the biological mechanisms that will enable him to respond effectively when the time of danger comes.

The word "adaptation" is treacherous. Among other things, it refers to the hereditary changes in genetic characters through which living things become better and better adapted to their environment from one generation to the next—the processes involved in Darwinian evolution. These genetic forces operate in man just as they do in other living things, and there is no doubt that they have been responsible during the course of millennia for the adaptive differentiation of human types in the various parts of the world. Clearly, the Negro possesses genetic traits which make him better adapted to unprotected life in the tropics than is the fair-skinned, blue-eyed Scandinavian.

Important as they are, these genetically controlled phenomena are not the adaptive processes which I have in mind here. Instead, I should like to emphasize the fact that during the past century mankind in the Western World has become adapted to very new conditions of its own making, and that even more drastic alterations in the ways of life will soon demand further adaptations. One century is a short time on the evolutionary scale for man, and for this reason genetic changes cannot possibly account for the adaptations that have occurred during the past few decades and that will be needed in the near future. On the other hand, and fortunately, each individual has in reserve an enormous range of potential adaptive resources that can be called into play under demanding circumstances. These adaptive potentialities have made it possible for millions of people to move in one generation from life on isolated farms to the tensions of Broadway and Forty-second Street, and we shall have to depend on similar mechanisms to survive the even more drastic and more rapid changes which are in the offing.

For many millennia mankind has moved forward and upward, even though in an erratic and halting manner. It has taken all sorts of disasters and upheavals in its stride, often deriving from painful experiences the stimulus for a more brilliant performance. We cannot create for our descendants a world free of stresses—nor, in my opinion, should we do it if we could. But, fortunately, there is no ground for believing that human nature has lost the resilience and creativeness that it has displayed heretofore. There is one aspect of the modern world, however, which has little historical precedent. In the past, most changes in the ways of life and in the physical environment occurred rather slowly, often taking several generations to reach the point of affecting large numbers of people. This slow rate of change permitted all sorts of progressive adjustments—at times through genetic mechanisms, more commonly through biological adaptations, and always through social devices. The fundamentally new aspect of the situation in our society is not that many changes are taking place, but rather that they occur so rapidly as to make orderly adjustments more difficult. Ambassador George Kennan has forcefully stated the importance of this new situation in our communities:

Wherever the past ceases to be the great and reliable reference book of human problems—wherever, above all, the experience of the father becomes irrelevant to the trials and searchings of the son—there the foundations of man's inner health and stability begin to crumble, insecurity and panic begin to take over, conduct becomes erratic and aggressive. These, unfortunately, are the marks of an era of rapid technological or social change.

Although Mr. Kennan's words refer to political and social issues, they are just as valid for the biological aspects of life—for all the reactions of body and soul to the ever changing environment.

The crucial consequence of this rapidity of change is that future generations will have to meet emergencies without benefit of their forebears' help or experience. The best thing that we can do for them, perhaps the only thing worth doing, is to create an atmosphere in which they will develop such nonspecialized adaptive powers that they can respond rapidly and effectively to all kinds of new and unexpected threats for which they cannot be specifically prepared. The objection may well be raised, of course, that these are idle words which correspond to a counsel of perfection without relevance to the facts of life. In part this is true, but in part only.

The reason that we know so little about how to make people develop their own adaptive powers is that modern civilization has not concerned itself with this problem. Everywhere in the world, and in the United States in particular, the trend has been toward controlling and modifying the external environment for the sake of human comfort, with total elimination of effort as an ideal. We do little, if anything, to train the body and soul to resist strains and stresses. But we devote an enormous amount of skill and foresight to conditioning our houses against heat and cold, avoiding contact with germs, making food available at all hours of the day, multiplying labor-saving devices, minimizing the effort of learning, and dulling even the slightest pain with drugs. Needless to say, I am not advocating a retreat from these practices which have made life so much easier although not necessarily very much happier. But I would urge that we emphasize more than we do now another approach to dealing with the external world—namely, the cultivation of the resources in human nature which make man potentially adaptable to a wide range of living conditions.

The field of adaptation has been so neglected and the knowledge concerning it is so scarce that to illustrate its manifestations with a trivial but concrete example might be useful. The illustration has to do with an experience common to all of us: our ability to endure hot and cold weather. When the outside temperature falls to 50° F. in September, everyone feels cold and additional clothing is is in order. In contrast, the same temperature in February evokes thoughts of summer and suggests a leisurely stroll. Clearly, the body becomes more or less adapted to the prevailing temperature as the season advances, and this ability to adapt has helped mankind to make its home over vast areas of the globe despite wide climatic differences. In view of this fact, it is probably unwise to avoid *completely* exposure to inclement conditions and thus prevent the human body from calling into action its temperature-adaptive mechanisms.

Let me acknowledge once more that little is known scientifically of the mechanisms involved in adaptation and that this ignorance makes it difficult to formulate rational courses of training. But it seems to me, nevertheless, that science can develop techniques compatible with civilized life and yet conducive to the development of general vigor and resistance. There is some indication, for example, that training for life in the tropics can be achieved by repeated short periods of activity in hot humid weather, and does not necessarily require *constant* exposure to unpleasant conditions. I have no doubt that techniques of adaptation could be developed for all sorts of strains and stresses if we were to devote our attention to the problem instead of relying exclusively on protection and escapism.

The considerations which have been presented in the preceding pages cannot yet influence the behavior of the practicing physician during the performance of his daily task in the treatment of disease. The moral responsibility of the physician in our society is to use all available resources for the succor of the sick and for the preservation of life, whatever the cost and the consequences. But the duty of the practicing physician toward his individual patient is only one aspect of medicine. Another aspect is made up of knowledge, practices, and points of view which bear on the welfare of the community as a whole, and on the future of mankind. And this has a large bearing, even though ill-defined because never discussed, on the formulation of a social philosophy of medicine.

In my opinion, it is meaningless and dangerous to encourage the illusion that health is a birthright of man, and that freedom from disease can be achieved by the use of drugs and by other medical procedures. Like political freedom, freedom from disease should not be regarded as a commodity to be distributed by science or government. It cannot be obtained passively from a physician or at the corner drugstore. Goethe's words apply here: "What you have inherited from your father, you must earn again or it will not be yours." Health can be earned only by a disciplined way of life.

It must be realized also that health and disease are concepts too complex and too subtle to be defined merely in gross physical terms. The meaning of these concepts is conditioned by the demands of the social environment and even more by the goals that the individual formulates for himself. Optimum performance imposes different health requirements on the plowman, the jet pilot, the philosopher. Thanks to medical science, we are in the fortunate situation of having today more than ever before the means and the knowledge to achieve the kind of health that we want, but this does not relieve us of responsibility. The truth is that the power at our disposal will be of no avail unless we work for the kind of health that we want, and this effort can be effective only if we define our individual and social goals and have the courage to make choices.

We must reconsider the wisdom of using individual longevity as the dominant criterion of social and medical ethics. We must be prepared to recognize that an excessive concern with security, with comfort, and with avoidance of pain and of effort has dangerous economic and biological implications—that such concern may, in fact, amount to social and racial suicide. I realize that any attempt to deal with these problems will involve painful conflicts with personal interests and with religious and moral convictions. Yet we have to formulate the problems in a forthright manner if we are to find their solutions. Unless we discover methods for producing a higher level of adaptive power in the individual and for preventing genetic deterioration of the race, the likely alternative is that more and more in the future we shall have to run frantically from one protective and palliative measure to another, trying to lengthen life at the cost of sacrificing its wholeness and many of its values.

Before closing, I must acknowledge that I have never taken care of the sick and am not a physician—a fact that has prevented me from apprehending with all their compelling force many of the human and practical aspects of medicine. Though fully aware of these deficiencies, I cannot refrain from quoting here a few lines from G. K. Chesterton's *Heretics,* brought to my attention by a humane physician who is also a scientist:

The mistake of all that medical talk lies in the very fact that it connects the idea of health with the idea of care. What has health to do with care? Health has to do with carelessness. In special and abnormal cases it is necessary to have care. . . . If we are doctors we are speaking to exceptionally sick men, and they ought to be told to be careful. But when we are sociologists we are addressing the normal, we are addressing humanity. And humanity ought to be told to be recklessness itself. For all the fundamental functions of a healthy man emphatically ought not to be performed with precaution or for precaution.

Chesterton was neither a scientist nor a physician, and as a sociologist he was prone to substitute brilliant paradox for logic and knowledge. Yet it seems to me that his flippant remarks help to quicken attention to an aspect of medicine that bids fair to become of increasing social importance in the future. Medical advances do not arise in a social vacuum. They are products of the sparks between the scientific knowledge of the time and the demands of the community. But what the community demands is determined to a large extent by publicity, apparent or hidden—in this case, by the implied promises of medical science. We must beware lest we create the illusion that health will be a birthright for all in the medical utopia, or a state to be reached passively by following the directives of physicians or by taking drugs bought at the corner store. In the real world of the future, as in the past, health will depend on a creative way of life, on the manner in which men respond to the unpredictable challenges that continue to arise from an ever changing environment.

The study of specific pathological problems requires the use of laboratory techniques and contributes to the advancement of laboratory knowledge. But the field of medicine transcends this kind of knowledge because it deals with man as a spiritual being and also with the future of the human race. Medical science is concerned not only with the control of individual diseases, but also with the long-range effects of its products on the total performance and happiness of the individual, on the social problems of the community, and on the adaptive powers of the race.

It must be realized, furthermore, that the attitudes of the physicians who are practicing the medicine of today and the efforts of the scientists who are creating the medicine of tomorrow are influenced—and to a large extent directed—by the beliefs and wishes of the culture to which they belong. If the social atmosphere puts a great premium on techniques and products designed to minimize effort, to relieve pain, and to increase the selfish enjoyment of today, these goals will be given priority by medical practitioners and research scientists. To a very large extent, this is what is happening today—in my opinion, at the sacrifice of higher values.

In medicine as in other social pursuits the long-range welfare of individuals and of the community often makes it advisable to forgo some of the immediate comforts and

This etching by Rembrandt,
in the possession of the Philadelphia Museum of Art,
is a portrait of an Amsterdam physician (1651).
It not only conveys an impression of compassionate understanding
of the plight of the patient, but also
symbolizes the need for the physician to apprehend
human problems in all their undefinable complexities.
Contrast the reflective, dreamy mood of this face
with the assurance of Descartes's expression
in the Hals portrait.

pleasures for the sake of the future. Overemphasis on the avoidance of effort and pain is an attitude fraught with dangers for the individual and even more for society; indeed, it amounts to social suicide.

Thus, medical philosophy at a high level transcends the problems posed by the care of the sick patient and must take into consideration the philosophical meaning of human existence. If we believe that the individual is but a link in a long chain of human adventure and that the continuity of human life is our collective responsibility, then it is wrong to jeopardize the future of the group for the sake of today's comfort. Medicine is one of the highest forms of social philosophy because it must look beyond the individual patient to mankind as a whole. The more effective it becomes through scientific knowledge, the more it must concern itself with the long-range consequences of its practices for future generations.

5

Illusions of Understanding

It was for reasons of convenience in presentation, and also to mark the four hundredth anniversary of Francis Bacon's birth, that I dealt first in this book with the applied aspects of science. In reality, however, it is not at all certain that science developed from utilitarian preoccupations, or that its practical importance is the main incentive for its growth in the modern world. The bourgeois credo that *Primum vivere, delude philosophari* has led to the assumption that the making of tools and the development of crafts preceded rational knowledge; but in fact primitive man seems to have been concerned with immortality and magic as much as with pottery and weapons.

There is clear evidence that the Neanderthal people, and before them men even more primitive, buried their dead with offerings. In Loren Eiseley's words, "It is the human gesture by which we know a man, though he looks out upon us under a brow reminiscent of the ape." Furthermore, as stated by Jacquetta Hawkes, "Materialism and preoccupation with technology has led to an exaggeration of the importance of tool-making as such; it may be far more significant that primitive man stuck feathers in his hair." No matter how primitive the human society, it is always organized around customs, myths, and beliefs that go beyond the requirements of a life of hunting and harvesting. Thus, it is probable that science evolved from nonutilitarian preoccupations of the mind as much as from craftsmanship; and it is certain that scientists in the ancient world were more concerned with philosophy than with technology.

The perennial fascination of Greek philosophy lies in its concern with the kind of knowledge that led man out of his brutish existence. Science would be just an instrument for comfort and power, not a cultural force, if it did not help man to transcend his animal origin. Whatever their selfish interests and their commitments to practical ends, most scientists cling to the faith—respected in the spirit even though often betrayed in action—that to work for knowledge and truth is the highest form of scientific duty. Ideally, and to a large extent actually, science is part of the collective effort for the humanization of mankind.

Modern science has been immensely successful in discovering facts and inventing techniques. As Francis Bacon had predicted in the *Novum Organum,* and as Ortega y Gasset scornfully emphasized in *The Revolt of the Masses,* even mediocre minds can make scientific discoveries. In contrast, it has proven far more difficult to recognize important relations between facts, and to convert the apparent chaos of reality into satisfying patterns. Only a few minds in each generation have been able to perceive the laws of the cosmos and to communicate them in a meaningful form to their less gifted fellow men. As to understanding the nature of the universe and of the human condition, it is questionable whether we have progressed much during the past two thousand years. The most that can be said, perhaps, is that science has helped mankind to dispel errors and overcome fears, and in addition has increased human awareness of the immensity and complexity of the cosmos.

True enough, there have been many times in history when man felt that he had acquired a body of knowledge and formulated a philosophy that provided a direct approach to real understanding. The illusion that this state had been reached was common among the Greeks of the classical period and among the medieval schoolmen. During the post-Newtonian era, also, many scientists and philosophers believed that the workings of the world could be so well understood that prediction of the distant future was possible. And around 1900 the chemist Marcelin Berthelot proudly claimed, "The world has no more mystery for us."

This statement symbolizes a faith that was prevalent especially among the French rationalists—namely, that science, and science alone, can solve the riddle of creation. In reality, however, there was much hesitation on this score even in France. Diderot, the standard-bearer of the Encyclopedists, was one of those who had moments of uncertainty, as one of his notes shows. "What do I perceive? Forms. And what besides? Forms. Of the substance I know nothing. We walk among shadows, ourselves shadows to ourselves and to others." As for the English scientists, many of them managed to retain throughout their lives an orthodox religious faith, apparently without bringing it into contact with a highly developed scientific sophistication. This was true not only of Newton but also of Joule, Faraday, Maxwell, and many others.

While orthodox religious faith in a scientist has been regarded by the pure rationalists as a form of intellectual abdication, the ability to suspend judgment on matters for which convincing information is not yet available might be regarded instead as a kind of deeper wisdom. The fact that the discoveries of modern physics have altered the meaning of the very concepts of objectivity and causality on which classical science was built may justify in the long run the conservative attitude of the English school. But even without appealing to the broader philosophical freedoms engendered by the relativity theory and the uncertainty principle, it would seem that understanding has not reached very deep, especially with regard to biological problems. What is considered understanding hardly ever, if ever, goes beyond familiarity with a certain group of phenomena which can be related in a convenient pattern of thought and from which limited predictions can be made. In most cases the illusion of understanding comes from a failure to examine the philosophical basis of one's professional knowledge.

In his "anniversary" address as President of the Royal Society in 1959, Sir Cyril Hinshelwood tried to relate certain recent biological discoveries to the concomitance of the outer world of nature and the inner world of conscious experience. He took his key from Heisenberg's famous remark that "The mathematical formulae no longer portray nature, but rather our knowledge of nature."

Professor Hinshelwood pointed out that what we know of life consists of observational data in which our consciousness is inextricably involved. For this reason we cannot validly disregard the part played by consciousness in the study of biological phenomena; the mind-matter relation cannot be ignored in total consideration of existence. In his words, "behaviouristic descriptions in general are models and abstractions ignoring data which so far from being trivial and irrelevant, are the only things which make us *inter alia* concern ourselves with science at all." In fact, it seems to me that the scientific attitude of modern biologists is conditioned to a very large extent by assumptions about the mind-matter problem which were made a few centuries ago and which are accepted as a basis of operation without concern for their validity, even by those who do not really believe in them.

During the early seventeenth century, as is well known, René Descartes asserted that the human body and the human soul are two separate entities and that the body is a machine which, therefore, can be studied as such. This was a convenient assumption and one which has proved extremely useful for certain kinds of scientific pursuits. Immediately following Descartes, scientists applied what they knew of mechanics to the body machine and found that its structure and functions were compatible with the knowledge derived from the study of lifeless systems. Then the chemists and the physicists engaged in similar studies and found that the phenomena associated with life obey at each step the same physicochemical laws that operate in the inanimate world. At the present time this approach to the study of the structure and functions of living things is culminating in the marvelous achievements of molecular biology.

The fact that Descartes's assumptions have led to such great scientific advances does not prove, however, that these assumptions are correct. There is no evidence what-

The original of this portrait of Descartes
by Frans Hals is in the Louvre.
The etching reproduced here shows a more sardonic smile
than is apparent in the original.
This helps to emphasize the contrast
between the intellectual arrogance of the rational scientist
confident of the power of his method
and the hesitating attitude
of the humane physician depicted by Rembrandt.

ever that the body and the mind are two separate entities, and, despite the triumphs of molecular biology, it has not yet been proven that the living body is only a machine and that life is merely a complex integration of known physicochemical forces. I realize that in raising this question I may seem to be reviving the vitalistic doctrine with all its false intellectual mysticism. But, in my opinion, I am doing nothing of the sort. I am only emphasizing that the machine view of living things is buried so deep in the modern subconscious that few scientists ever try to bring it to the surface to examine its significance in the bright light of critical knowledge. And I believe that the acceptance of an oversimplified mechanistic theory of life has narrowed considerably the front of progress in biological sciences.

I shall introduce the discussion of this topic by looking at conventional machine biology in its most favorable light and pointing to what I regard as one of the most important contributions of modern science to philosophy— namely, the doctrine of the chemical unity of life.

It has been found that all living things—whatever their size and whether they be man, animals, plants, or microbes—possess many physicochemical characteristics in common. In particular, all depend upon the same fundamental reactions for their supply of energy; all synthesize proteins of approximately the same amino acid composition for structural purposes and for enzymatic activity; all transfer their hereditary endowment from cell to cell through the agency of submicroscopic particles consisting largely of nucleic acids. These physicochemical similarities provide, of course, spectacular confirmatory evidence for the theory of evolution. They greatly increase the likelihood that all living forms studied so far had a common origin. Starting from a single point of genesis, organisms have progressively differentiated, but have retained the most fundamental characteristics associated with life at its beginnings.

The chemical unity of life is compatible with two different working hypotheses. One is that some unknown principle runs like a continuous thread through all living forms and governs the organization of their physicochemical properties. The other is that there is no such principle and that forces known to operate in the inanimate world are sufficient to account for all characteristics of life. The second hypothesis is, of course, a restatement of the machine analogy that has prevailed since Descartes's time —without mention of the soul that he had associated with the body. Of the two hypotheses, it is now the more generally accepted because it is economical of thought and also because biological phenomena have been found to go hand in hand with physicochemical reactions. In reality, however, the fact that these phenomena do not contradict known physicochemical laws is not sufficient evidence to prove that life is merely an expression of these laws. Correlation and lack of contradiction could be compatible with other theories of life.

It is apparent, furthermore, that living processes differ in many fundamental ways from what is known of the inanimate world. For example, whereas an increase in entropy is expected in all inert systems, living things produce and maintain order from the components of the inanimate world, and this property disappears the very moment life ceases. Or again, living things possess the property of duplicating themselves and at the same time of undergoing changes which they can also replicate; nothing of the sort is known in the inanimate world. As for the mind, whatever its nature and whether or not it is a separate entity, the constellation of characteristics that it symbolizes confers on life powers of selection and of action not recognized in things not living. Let it be emphasized once more that these obvious differences are mentioned here not to uphold the doctrine of vitalism in its worn-out form, but rather to illustrate how difficult it is to equate life with *known* mechanisms. In fact, to equate life with matter and its laws requires something beyond scientific imagination: it demands the *a priori* faith that living things *are* nothing but physicochemical machines.

The greatest merit of this faith is that it has engendered beautiful experimental work. During recent years it has also stimulated several chemical hypotheses to account for the origin of life. Indeed, actual programs of experimentation to put these hypotheses to the test are presently under way or are being planned. To illustrate their intellectual boldness, it will suffice to quote here from an authoritative paper on this theme by S. L. Miller and H. C. Urey, published in the July 31, 1959, issue of *Science*.

The major problems remaining for an understanding of the *origin of life* are (i) the synthesis of peptides, (ii) the synthesis of purines and pyrimidines, (iii) a mechanism by which "high-energy" phosphate or other types of bonds could be synthesized continuously, (iv) the synthesis of nucleotides and polynucleotides, (v) the synthesis of polypeptides with catalytic activity (enzymes), and (vi) the development of polynucleotides and the associated enzymes which are *capable of self-duplication.* [Italics mine.]

This statement translates Descartes's assertion into modern chemical language and takes it for granted that the apparent differences between animate and inanimate matter are merely the consequences of a somewhat greater chemical complexity in the former. Whether or not a crucial experiment can be designed on this basis for the production of life *de novo* out of inert chemicals is a question that has to be answered "on faith" until the experiment has been successfully performed. But even if it proves possible to create a self-duplicating system by assembling the proper kinds of nucleic acids, proteins, and other organic molecules, the question of the origin of life will not be decided thereby. An analogy—admittedly superficial and unconvincing, as all analogies are—may help to convey the nature of the difficulties that I see in the interpretation of any experiment designed to create life *de novo*.

Let us assume an investigator who knows everything about the construction and operations of modern electric engines, but nothing of the scientific history of their development. This investigator can obtain and assemble all the parts required for constructing the engine, and, by supplying energy, can get it to function. Stretching the meaning of the terms, this feat might be regarded as "creating" a functioning electrical engine—but certainly it would prøvide no information as to the "origin" of the engine or about the nature of the forces involved in its operation. What the investigator would have done would be to have confirmed that the available technical information about electric engines could be used to produce an expected result.

Likewise, assembling a self-duplicating system from complex molecules known to be present in systems which are living and therefore self-duplicating would prove that man—himself a living thing—can reconstitute some of the mechanisms which he has found to be associated with the living objects that he knows. Important as the achievement would be, it would not settle the question of the *ori-*

gin of life. In this regard it is interesting to note that most of the authors who have formulated hypotheses to explain how life *first* emerged from matter have assumed that the initial reactions which could be regarded as having some similarity to the living process occurred, not in nucleic acids or proteins, but in simpler and more primitive molecules. Only progressively in the course of æons would the machinery of life have become transformed into the very complex apparatus that we know today. If it has any meaning at all in the terms of our present scientific philosophy—which is questionable—the origin of life must therefore be sought in processes unlike those found in the highly evolved living things that have survived and that we know.

Life has many properties which are at least as interesting, and as puzzling, as the ability of living things to reproduce themselves. One such property is the dynamic stability of biological systems, the fact that they retain their morphological identity and other essential characteristics while being most of the time in a state of vigorous metabolic flux.

In addition to the aspects of life which are for the time being best described in physicochemical terms, there are others which are still the distinct province of the classical biologist—a fact which does not make them less important, less modern, or less "scientific." A few familiar examples will suffice to illustrate that the understanding of life will remain an illusion as long as one cannot account for the origin of reactions and of types of behavior found only in living things.

One of the most puzzling aspects of living processes is that among known organisms capable of independent life, those which are probably the most primitive—namely, the bacteria—appear to be the most versatile and the most completely equipped from the biochemical point of view. Thus, certain species of bacteria can utilize very primitive sources of energy and can synthesize all their organic constituents from carbon dioxide and minerals. From what can be observed, the main trend of evolution seems to have been not the development of new chemical activities, but rather the coordination of existing processes in time and in space.

During embryonic development the existing structures and functions are so designed as to contribute not only to the immediate needs of the embryo, but also to the future development of the fetus. More generally speaking, the functional and morphological changes that take place at any stage in the development of the embryo or fetus anticipate the future functional and morphological requirements of the fully developed organism and of its progeny. To state that this wonderful arrangement is the outcome of evolution is a descriptive account of what has happened, but provides no understanding of how and why. It may be worth mentioning here that the original meaning of the word "organ" is "working tool." This etymology symbolizes the fact that describing an organ and studying it by the techniques of all known sciences cannot so far provide a complete understanding of it. The organ acquires its full significance only when it functions and when its performance is integrated in the life of the organism.

Not only is it impossible to really understand a structure apart from its role in the whole organism, but in addition there exist in living things potentialities that become manifest only under very special conditions. The phenomena of symbiotic relationships reveal how these potentialities can generate reactions which have large creative effects—as illustrated in the two following examples.

Small and frail as they appear to the untrained eye, lichens are in many ways among the most successful living things. They occur in many intriguing forms, with a wide range of colors, on the bark of trees, on rocks, and on waste lands. One of their characteristics is their ability to become established and to prosper under the most inimical conditions, even in places where life appears all but impossible. They are the first living things that develop on bare rocks, and they are abundant even in the most desolate areas of the Antarctic.

What makes lichens so important for our discussion is that they are made up of two different microorganisms living in intimate association. In fact, the word "symbiosis," which means "life together," was invented to refer to the kind of biological association found in lichens. Each lichen is the symbiotic summation of one species of alga and one species of fungus, the two organisms being so intimately interwoven that it is extremely difficult to separate them. There is no doubt that the alga and the fungus supplement each other nutritionally when they are associated in the form of a lichen. The alga produces chlorophyll and is therefore capable of manufacturing carbohydrates from the carbon dioxide of the air by photosynthesis. The fungus feeds on these carbohydrates, and in exchange it extracts from the environment certain minerals which are used by the alga; furthermore, the fungus acts as a reservoir of moisture. Although these interrelationships have not yet been worked out in detail, it is clear that they are of advantage to both members of

Lichens were long thought to be ordinary plants.
They are in reality the composite expression
of an alga (A) and a fungus (B)
living in symbiotic association.
This association results in creative effects
that are never produced by the fungus
or the alga living alone.

the partnership. In any case, it is certain that lichens can multiply where nothing else will grow and can survive under conditions that seem incompatible with the survival of other living things.

The root nodules of leguminous plants constitute another striking example of symbiosis. These nodules are produced by the response of the plant to the presence of special kinds of bacteria called *Rhizobium*. Although these bacteria become established on the rootlets and multiply in certain cells, normally they do not invade the rest of the plant. Leguminous plants can exist without root nodules in the absence of *Rhizobium* bacteria, but usually grow better when associated with them, particularly in soils deficient in nitrogen. The root nodule bacteria can fix gaseous nitrogen from the air and convert it into organic nitrogenous compounds that can be utilized by the plant. In exchange, the plant supplies the bacteria with other nutrients necessary for their growth. The partnership, therefore, is beneficial to both partners; it constitutes a true symbiosis between plants and bacteria.

The nutritional aspects of symbiosis have been the most extensively studied, because nutrition is a factor of such obvious importance for survival and for growth and, moreover, one that can be readily analyzed. It is certain, however, that symbiosis often has other effects of greater biological interest. Thus, lichens exhibit complex morphological structures and synthesize peculiar organic acids and pigments that neither the alga nor the fungus can produce alone. Looking at the delicately shaped and bright red structures of the common lichen known as "British soldier," it is difficult to believe that such startlingly beautiful forms can result from the association of a microscopic alga and a microscopic fungus, both of them inconspicuous. Nor would it have been possible to predict from the known characters of its two components that the lichen could synthesize the peculiar chemical substances that it produces, or exhibit such great ability to survive heat, cold, or dryness.

The association of the leguminous plants with *Rhizobium* provides another astonishing example of the creative effects of symbiosis. The root nodules contain a red pigment which is almost identical with the hemoglobin of red blood cells! This form of hemoglobin cannot be produced by the bacterium or by the plant alone, nor, for that matter, is it known to occur anywhere else in the plant world. It seems to be formed only in the plant cells which harbor the *Rhizobium*. Thus, it is clear that symbiosis is more than an additive association; it is a creative force which can result in the production of unpredictable new structures, functions, and properties.

One could quote many other examples of biological phenomena not predictable from the knowledge of forces usually studied. Such are the biological cycles which reveal dramatically how intimately all organisms are related to the diurnal or annual movements of the earth or to the phases of the moon; the profound effects of latitude, as well as of winds and weather, on physical well-being and on behavior; also perhaps some of the odd claims of parapsychical research. I mention these facts not as an appeal to magic but merely to emphasize that there exist in heaven and earth more things than appear in present-day scientific philosophies. To acknowledge ignorance of these important matters is, in my opinion, not a retreat from reason but instead the most constructive way to broaden and deepen the scientific approach to the understanding of life.

Increasingly during recent decades the study of biological problems has been influenced by two large assumptions which at first sight appear to be based on hard-boiled scientific common sense, but in reality are still *sub judice*. One is that life can be understood only by analyzing the mechanisms linking the molecular and the animate worlds; the other is that the arrow of influence between these two worlds points in only one direction, from the molecular lifeless components to the more complex patterns of organization found in living things. These two assumptions have been immensely fruitful because they have encouraged investigators to break down phenomena and structures into smaller and simpler components, ultimately to be described in terms of identifiable chemical forces and substances. Moreover, they provide the easiest and safest approach to biology. They free the scientist from the need to engage in soul searching about the philosophical meaning of life, since in the final analysis they equate living processes with the reactions of inanimate matter. Finally, they permit an endless series of laboratory operations, because to disintegrate and analyze is far easier than to build up complex functioning organisms or even to investigate them as a whole. In the words of Professor Homer W. Smith,

I would define mechanism, as we use the word today, as designating the belief that all the activities of the living organism are ultimately to be explained in terms of its component molecular parts. This was Descartes' greatest contribution to philosophy. . . . Abandon Cartesian mechanism and you will close up every scientific biological laboratory in the world at once, you will turn back the clock by three full centuries.

It is likely, however, that if the analytical breakdown of living things into simpler and ever simpler components is not supplemented by a more synthetic approach, it will lead the biologist into areas of knowledge concerned not with the essential characteristics of life, but with a few selected phenomena which happen to be associated with living processes. To accept this limitation is an attitude of intellectual security and may be the better part of wisdom, but it denies scientists the chance to gain deeper insight into larger biological realities. As a contrast to the unphilosophical endless accretion of "scientific" facts concerning living *matter,* it is stimulating to rediscover in Aristotle's writing the entrancing throb of life. Darwin had this experience on reading William Ogle's translation of *The Parts of Animals.* "I had not the most remote notion what a wonderful man he was," wrote Darwin. "Linnaeus and Cuvier have been my two gods, though in very different ways, but they were mere schoolboys to old Aristotle."

For those who believe that scientific biology is synonymous with the more and more refined study of well-defined, isolated reactions, it is chastening to remember that the greatest biological generalizations were not reached by this analytical method. One of the few universal principles in biology—namely, the Darwinian concept of evolution through the natural selection of random hereditary fluctuations—emerged not from a study of the primary units of Mendelian or biochemical genetics, but from inspired guesses based on a sort of Gestalt awareness of complex relationships in natural situations. The modern scientific techniques have served merely to verify the theory and to elaborate its details.

Even with regard to practical applications, biology has made some of its most important advances by techniques

Ever since Charles Darwin (1809-1882),
evolutionary concepts have conditioned all scientific thought.
As pointed out by Theodosius Dobzhansky,
in a criticism of Chapter 5 in this book,
"Life would be indeed an utter puzzle,
if it were not that it took something between two and five billion years
to arrive at the wonderful state
in which we now observe the living things.
Evolution is a mechanism which makes probable
what otherwise would be in the highest degree improbable."

that did not involve the reduction of phenomena into smaller and smaller elements. Many illustrations could be quoted in support of this point of view. Thus, the modern ecological doctrine that mixed complex biota are more stable than biota consisting of only few species could not have been developed except from the study of the fate of species and their interplay under natural conditions. Or, again, the immunological theory, which is playing such an immense role in modern medicine, developed at first, in the absence of any chemical knowledge of antibodies or immune reactions, from the observed fact that living things respond to almost any kind of stimulus by a set of reactions more or less specifically directed against the stimulus. Indeed, a useful biological philosophy could be formulated on the basis of a Le Châtelian type of auto regulation—a negative feedback of organisms in response to their environment—even in ignorance of the precise mechanisms involved.

Conditioned reflexes constitute another class of phenomena whose meaning and very existence are expressions of the responses made by the organism as a whole to its total environment. It is of interest to note in passing how the direction of physiological science can be influenced by the current scientific philosophies in different national groups. In general, American textbooks of physiology begin with a discussion of the physicochemical and mechanical principles which govern the cell and the cardiovascular system. In contrast, a recent authoritative Soviet text begins with the Pavlovian concept of holistic physiology as the basis for behavioral analysis. The pervading influence of the physicochemical point of view in Western biology is well expressed in the following passage taken from an American author's review of this Russian text in a recent issue of *Science:*

The conditioned reflex technique is a means of precise description and prediction rather than of understanding. It is a symbolic language used to describe behavioral patterns without recourse to theory. On the one hand it does not lead to the joining of physiology with physics and chemistry, which ultimately afford the basis for explanation. On the other hand it does not require the use of tenuous behavioral concepts such as motivation, reward, punishment, emotion, or memory in order to describe complex patterns of somatic and visceral activity.

The study of physiology involves the attempt to answer the question about an organ, "How does it work?" To a Soviet physiologist this appears to mean, "What is the observed correlation (law) of events?"; whereas to his Western counterpart it more often means, "What is the major premise (law) by means of which one event can be said to follow logically from another?"

The contrast in attitude between American and Russian workers in neurophysiology is apparent in the fact that, whereas the former concentrate their efforts on the physicochemical determinants of neural activity, the latter are chiefly concerned with its manifestation in the living individual.

At the cost of being repetitious, I shall elaborate once more on the points of view expressed in the preceding pages, in order to make clear that they do not express a nihilistic attitude toward the scientific study of biological problems. Life is more than a self-replication of nucleic acid and protein molecules, supplemented now and then by a few mutational changes. It is more than the utilization of chemical energy for the synthesis and turnover of organic materials. Life is a creative process elaborating and maintaining order out of the randomness of matter, endlessly generating new and unexpected structures and properties by building up associations which qualitatively transcend their constituent parts.

In his enlightening book *The Edge of Objectivity,* Professor Charles C. Gillispie has brilliantly defended the view that the most productive philosophy in Western science has been atomicism, the reduction of force, matter, and reactions into their ultimate constituents. Concern with being rather than with becoming, Professor Gillispie claims, has proven in final analysis the most fruitful attitude for the discovery of facts and of laws. While the history of physics and chemistry provides some support for this contention, I believe that the evidence concerning biology is far less convincing. In my opinion, exclusive preoccupation with "being" leaves out of biology the very phenomena which differentiate the animate from the inanimate world. The fact that words like "becoming" or "emergent evolution" cannot yet be defined in meaningful, operational terms does not deprive them of scientific importance. Rather their vagueness symbolizes a lack of understanding of biological phenomena—an ignorance increased by our cultural tendency to shy away from the peculiarities which make life so obviously different from matter. For this reason it is unwarranted, indeed unbiological to limit the study of problems of life to the analysis of fragments or reactions isolated from the organism by techniques which first destroy or at best inactivate life. We must strive for the development of more sophisticated techniques dealing with the creative aspects of living processes. It might be rewarding to return for a while to the Aristotelian view that the living organism cannot be understood except as an organized pattern of responses to the environment.

Living things cannot be differentiated from the inanimate world in terms of structures and properties. Their uniqueness resides in the fact that their functions and behavior are determined by their present environment, their past, and their future. This view is compatible with but goes far beyond mechanical behaviorism. It takes into consideration the ability of the living organism to exhibit selective responses determined by its past and conditioned by awareness of the future, a property as yet mysterious but real nevertheless. Life is historical, and man is its most advanced expression in that his activities reflect his past and are turned toward his future, that they are more intensely concerned with his memories and his goals. This does not mean that the study of man's nature is outside the scope of science. Far from it. The limitations to the understanding of man are not different from those inherent in the search for any kind of knowledge, whether it be knowledge of mind or of matter. If there are any limitations, they arise from the strictures imposed on science by unproven assumptions. There was a time when it was considered simplest, and indeed most elegant, to explain the universe in terms of heavenly bodies moving in perfect circles around the earth. Similarly, the desire to fit the phenomena of life into the simple pattern of the known laws of inanimate matter is attractive and economical of thought, but it conflicts with reality. By selecting the facts that fit the body-machine concept one will discover the physicochemical laws that govern some of the mechanical operations of living things, but one will leave out of consideration the creativeness of life, and the values of man.

In a recent essay Professor Theodosius Dobzhansky pointed out that the evolution of man corresponds to "a natural process that has *transcended itself*. Only once before, when life originated out of inorganic matter, has there occurred a comparable event." Professor Dobzhansky urged also that the time has come for man "to replace the blind force of natural selection by *conscious* direction based on his knowledge of nature and on his *values*." (Italics mine.) The statement that natural processes have had to *transcend* themselves to produce first life, then consciousness and human values, out of inorganic matter acknowledges in fact that evolutionary theory in its present form does not account for the emergence of man from the inanimate world. (For a criticism of this view, see page 117.)

It seems to me that a return to the Aristotelian philosophy, far from being a retreat, would enlarge the scope of the biological sciences. Biology will run dry unless it becomes more receptive than it is presently to unsuspected phenomena, unpredictable on the basis of what is already known. Science does not progress only by inductive, analytical knowledge. The imaginative speculations of the mind come first, the verification and the analytic breakdown come only later. And imagination depends upon a state of emotional and intellectual freedom which makes the mind receptive to the impressions that it receives from the world in its confusing, overpowering, but enriching totality. We must try to experience again the receptivity of the young ages of science when it was socially acceptable to marvel. What Baudelaire said of art applies equally well to science: "Genius is youth recaptured." More prosaically, I believe that in most cases the creative scientific act comes before the operations which lead to the establishment of truth; together they make science.

Many great experimenters in all fields of science have described how their ideas were determined in large part by unanalytical, visionary perceptions. Likewise, history shows that most specific scientific theories have emerged and have been formulated gradually from crude intuitive sketches. In this light, the first steps in the recognition of patterns or in the development of new concepts are more akin to artistic awareness than to what is commonly regarded as the "scientific method." I have purposefully used vague terms such as "visionary perceptions" and "artistic awareness" with the knowledge that this terminology will cause accusations of antiscientific and even antiintellectual mysticism. In reality, however, I do not believe that my attitude is based on a naïve acceptance of intuition as a sort of second-class revelation. Instead, it is determined by the belief that scientific questions have their origin deep in human consciousness, often below the analytical level. They constitute specialized restatements of large questions that philosophers formulated long before scientists began to work on their determinism, questions which have indeed preoccupied men ever since they began to think—even before the beginnings of formal philosophy. Many ancient myths are the first statements in a symbolic form of abstract themes not yet formulated in philosophical or scientific terms.

The most complex scientific concepts can often be recognized in ancient writings—in a very primitive form, but expressed in arresting images, such as the "torch of life," which symbolize profound truths of nature. For example, the modern biochemist who demonstrates that the components of the tissues are constantly renewed in the body—in a state of dynamic equilibrium, as the expression goes—seems to provide evidence for the very ancient belief that through all the fluxes of matter and energy the organism stubbornly maintains its own particular individuality. "Man is like a fountain," said Heraclitus, "always the same but never the same water."

Just as individual scientists have expressed many different and apparently opposite views concerning the ways of nature, so have philosophers. While Heraclitus taught that everything is constantly in flux, Ecclesiastes lamented that nothing is new under the sun. If scientists often appear to contradict one another, it is because complementary views are needed to express the multiple aspects of reality. Thus, the theory of dynamic equilibrium, which was the last word in biochemical sophistication when first enunciated three decades ago, is now being questioned again by a new generation of biochemists. In a very recent series of lectures an eminent biochemist suggested that life might reside in the stability and continuity of nonliving macromolecules within the cells, rather than in the transformation of components which undergo rapid turnover. According to this biochemist, in other words, the essence of life might be found in the Cartesian concept of *being* rather than in the concept of *becoming* which had its origin in Heraclitus of Ephesus and which Hegel popularized. Universal instability of constituents seems to be compatible with a stability and even monotony of organized life.

Ancient myths and literary images remain meaningful to us probably because they symbolize problems crucial for mankind which are still unsolved. In fact, it can be said that the problems which modern scientists regard as fundamental are precisely those which have constituted the subject matter of philosophy since the beginning of time. For a large body of scientists the great problem is the nature of the cosmos, the origin in time of its present ordering, the laws (cyclic or otherwise) of its evolution. What is the cosmos made of—waves or particles? And what are the mechanisms of their metamorphoses? For other scientists the final questions of science as of philosophy are not about matter and energy, but about the nature of life in a universe where lifelessness is the rule, life the puzzling exception. How do living things differ from inanimate matter? How did they originate? And can life be created *de novo?* Is man qualitatively different from the rest of the living world or merely a higher specimen in its evolution, the paragon of animals?

The science of the mind also is reformulating some very old preoccupations. The modern psychologist is close to Aristotle when he recognizes that the human personality can be divided into three parts: a basic, primeval self—biological substratum or id; a derivative social self, transmitted by man's culture; an ideal self, the superego. Are psychologists talking science or philosophy when they state that one is born with the first self, one is born into the social self, one must be reborn if one is to achieve the third self? Even the problems of ethics and social organization emerge from the observations and experiments of biologists in a form that would look familiar to ancient philosophers. The scientist who studies social insects—bees, termites, or ants—cannot help asking himself whether society should be organized for the sake of the individual or of the group, and whether too efficient an organization may not correspond to a static view of life incompatible with the development of mankind and the urges of the human spirit.

It is a remarkable fact that, in their efforts to formulate the central problems of modern science, experimenters as well as theoreticians tend to phrase their questions in the form of alternatives which, formally at least, resemble those considered by ancient philosophers. A large part of the history of science, for example, appears to continue the debate begun several millennia ago between the proponents of the continuum and the atomicists. As Professor Gillispie emphasizes, the duality between these two points of view has provided a dialectic for much of science ever since the beginning of the debate in Greece.

Among biologists the debate took its most lively form in the long-lasting controversies concerning the differentiation and the evolution of living forms. Goethe and Lamarck are among the most celebrated standard-bearers for those who want to find in nature unity and unbroken continuity. In Leibniz's words, *"Natura non facit saltus."* But ever since Democritus most scientists have found it easier to see the phenomena of the natural world as a multiplicity of discrete events. As is well known, the gene theory today dominates evolutionary concepts and has brought much of modern biology within the fold of the school of "elementary" particles. It is not impossible, however, that as the gene concept becomes more sophisticated, the changes undergone by living things may again appear progressive and once more fit the Heraclitean flux and progress better than Democritean atomicism.

Mathematical, physical, and chemical theories offer, of course, many illustrations of the endless debate between atomicists and partisans of the continuum. The theoretical physicists seem to have occupied the center of the stage during the past half-century, and their debates concerning the relative merits of the corpuscular and wave theories have spread from discussions about the nature of light to other aspects of matter and energy. Here again the point of view identified with Democritus seems in accord with the discovery of elementary particles and with the fact that the quantum theory is extending atomicism almost indefinitely. But men still find it difficult to divide time into discrete instants, and space into discrete points.

Only sophisticated physicists could discuss usefully these profound problems, but even the outsider cannot help noticing how mathematicians and physicists are prone to use words and to take points of view which present some formal analogy with those used by laymen and classical philosophers. Heraclitus saw fire as the basic stuff of all creation, and, similarly, the modern physicist regards energy as the substratum from which all matter is made. The matter of Aristotle which is mere *"potentia"* can be compared to the concept of energy which gets into "actuality" when the elementary particle is created.

There may be some profound meaning in these similarities in the language of modern scientists and of ancient philosophers. After all, ordinary language, vague as it is, evolved in immediate connection with reality, and the concepts that it tries to communicate are reality itself. As stated by Mr. Curtis A. Wilson of St. John's College,

There is a rhythm and structure in ordinary speech which forms the seed from which all the larger unifications grow—a Greek tragedy, a Mozart concerto, or the Einstein-Minkowski geometrization of space-time. Myth and logic, poetry and the sciences, alike develop out of the semantic and syntactic potentialities of everyday speech.

Because of this fact, the common language may be more adaptable to the expansion of knowledge than are precise scientific terms which correspond to limited and selected groups of phenomena rather than to reality as a whole. Furthermore, human thought evolved in direct associations with nature. Our thought patterns are based not so much on recently discovered evidence as on perceptions long experienced and on facts long known which may have influenced the formation of human intellect. Indeed, the astonishing ability of the human brain to guess certain workings of the universe suggests that, in a limited measure at least, the brain mirrors some of its very patterns. In this light, it becomes less surprising that ancient philosophers perceived the central core of many of the problems of modern science and that their formulation of these problems presents great similarities to our own. The search for viable truths and all the other worldly preoccupations of science are but part of the high tradition of religion and philosophy from whence most knowledge originally sprang.

6

The Dehumanization

of the Scientist

It is symbolic of the responsiveness of scientists to their social environment that Professor DuBridge devoted the first series of Pegram Lectures to space science. G. B. Pegram had "the conviction that the results of science can be made to serve the needs and the hopes of mankind." While practically all scientists share his conviction, for most of them it is probably more in the nature of a religious faith accepted a priori than a well-defined doctrine based upon logic or factual evidence. Professor DuBridge's lectures on *Introduction to Space* illustrate well that the relation of science to human needs and hopes is often indirect and tenuous.

The immensely exciting quality of space science in all its engineering and theoretical aspects is obvious to everyone, to the crudest barbarian and the village fool as well as to the thoughtful scholar. But as Professor DuBridge makes clear, it is far less certain that space exploration will soon or ever serve the "needs" of mankind—at least those needs of which ordinary men and women are aware. As to the "hopes" to be served by space science, they are very great indeed, and some of them are within reach; but on the whole they concern our knowledge of the universe rather than the more common aspirations of the body and of the heart. Yet, despite these limitations, it cannot be questioned that space vehicles are crowded with the most ancient dreams of mankind. Not only do the skies and the remote worlds give to man the promise of new experiences; even more important is the fact that they provide him with the illusion that he can escape from the earth— just as dreams of the future or of the past provide him with an escape from the present.

Imaginary travels to the outer world are probably as old as human imagination. In *Moon Travellers* Peter Leighton has presented a long list of them, beginning with early Greek writers and continuing with Plutarch, Cyrano de Bergerac, Jonathan Swift, Jules Verne, and H. G. Wells. Cyrano de Bergerac deserves special mention among writers who tried to translate into fiction the human longing to escape from earthly bondage. He lived from 1619 to 1655 and is remembered because of his large nose, for the honor of which he successfully fought countless duels; his romantic life was made famous by Edmond Rostand's play a generation ago. In addition to his facial peculiarity, this tempestuous *cadet de Gascogne*—of the kind immortalized in Alexandre Dumas's *The Three Musketeers*—had the distinction of being a philosopher who had studied physics with Gassendi and who was in revolt against the religious and social mores of his time. As a form of protest against the world in which it was his fate to live, he imagined that he could transport himself to the moon and other astral bodies and there find societies agreeable to his tastes. He described these imaginary experiences in two books which were published between 1657 and 1662 under the common title *L'Autre Monde — Voyages aux empires de la lune et du soleil*. These two books, now three hundred years old, are still of interest by virtue of Cyrano's vivid imagination. His account encompasses the use of heated air to ascend into space, the corpuscular theory of matter, the description of a phonograph and its use for teaching purposes, the practice of eugenics on the moon, the teaching that physicians must keep people well rather than treat diseases, etc., etc. These wonderful tales, which enjoyed success for almost two centuries, now remain undisturbed under the dust of library shelves; but they live nevertheless in all the books on space travel that they have engendered—including Jules Verne's *From the Earth to the Moon*, H. G. Wells's *A Modern Utopia*, and the science fiction of our day.

My purpose in evoking here the picturesque memory of Cyrano de Bergerac is not so much to give him the credit that he deserves for his early contribution to space travel literature as to illustrate his use of science to express an escapist mood. Cyrano de Bergerac was a very gifted and well-educated man, but a social failure and unhappy. His travels to the moon and to the sun, and the pleasant societies he imagined there, provided him with relief from his trials on this earth. He was well aware of his escapist attitude, as is revealed by the following verses that accompany one of his four known portraits:

> La terre me fut importune
> Je pris mon essor vers les cieux
> J'y vis le Soleil et la Lune
> Et maintenant j'y vois les Dieux.

Today space science is a reality and there are many good orthodox scientific reasons for its enormous appeal to the public, but it is certain that most of this popular interest has no scientific basis. There is, of course, the sheer excitement aroused by anything spectacular and new; and even more, perhaps, there is the desire to escape—if only in imagination—from the difficulties of everyday life and also from its boredom. Cyrano himself spoke of the Empire of the Moon as a place where "even imagination is completely free." And now science provides for the imaginations of laymen the richest material for daydreaming that all the problems of the world can be solved by magic.

That "science will find a way out" is a dangerous illusion because it serves as an excuse for intellectual laziness and apathy. I shall try to show that this attitude, which corresponds in reality to a deterioration of public interest in the intellectual aspects of science, originates in part from a change of ideals among the scientists themselves.

Until Bacon's time the motivation of scientists was either plain curiosity or the philosophical urge to understand the world; the practical problems of life were hardly ever mentioned as a justification for their efforts. This does not mean that practical matters did not orient and influence somewhat the activities of scientists. It is obvious that in the past as today what scientists did was necessarily conditioned by the techniques at their disposal and by the preoccupations of their times. But the enormous gap that existed until 1800 between the large amount of theoretical knowledge and the paucity of applications derived from it bears witness to the fact that very few of the ancient scientists focused their efforts on practical issues. One example must suffice to illustrate how profoundly the Industrial Revolution changed in this regard the professional outlook of the scientific community.

William Thompson (Lord Kelvin) had proved himself

a most gifted theoretical investigator during the early part of his life. Before the age of thirty-three he had published studies which constituted the foundation of thermodynamics, provided Maxwell with the mathematical clues to the electromagnetic theory of light and led Hertz to the discovery of radio oscillations. But instead of pursuing the large theoretical implications of his discoveries, Thompson soon shifted his efforts to technological developments and became the first great scientist to organize a laboratory devoted to industrial research. The change in scientific ideals that he symbolizes constitutes one of the most important characteristics of the nineteenth century. It has had such far-ranging consequences that we must consider in greater detail the forces involved in the conflict between the philosophy that "the true scientist has elected to know, not to do" and the other attitude, more common today, that the true role of science is to be, in I. Bernard Cohen's words, the "servant of man."

"All men possess by nature the desire to know." These words, with which Aristotle began his *Metaphysics,* correspond to one of the most characteristic human traits, even though the thirst for knowledge may not be as universal as Aristotle would have us believe. Spontaneously all children observe, explore, ask for explanations, and invent one if their request is not satisfied. The fact that all human beings thus begin as embryonic scientists focuses attention on two related problems concerning science and scientists. One has to do with the factors that cause most adults to lose the intellectual curiosity which was theirs as children. The other concerns the mysterious reasons which drive a few individuals to devote themselves to theoretical science. As Einstein pointed out, what is surprising is not that

China and India did not create experimental science, but rather that Europe did. It is plain that, whereas curiosity is natural to man, the discipline of science is not spontaneous and demands a painful effort.

All the great passions and occupations of mankind have been analyzed in the literature of all times, but only during recent decades has the scientist attracted the interest of students of human nature. Increasingly, however, his motivations and the qualities that make for his professional success are being depicted as those of soldiers, farmers, or bankers have been in the past. From Maurice Arthus's once famous series of essays on *The Natural History of a Scientist,* through Sinclair Lewis's *Arrowsmith,* to C. P. Snow's novels, many books have tried to throw light on the life of science and on the idiosyncrasies of scientists. Imagination; informed and disciplined curiosity; an interest in form, in forces, and in patterns; the ability to engage in abstract contemplation as well as in objective observation; delight in intellectual effort; resourcefulness; a taste for power and a desire to benefit the world—all these traits in various combinations have been reported to be common among scientists. Obviously, however, many laymen also exhibit these motivating forces or qualities, and, furthermore, it has not been shown that scientists lack any of the attributes found in other professions. In brief, scientists do not seem to differ in any consistent manner from other groups of human beings.

Since no trait peculiar to scientists has yet been recognized, it seems best to assume that at all times and in all places approximately the same percentage of men has been endowed with the qualities which make for effective scientific performance, but that conditions for the emergence of scientific vocations have not always been equally favorable. It would seem, in other words, that the accidents of individual life and, even more, the demands of the environment, as well as the facilities that it offers, are the most important factors in determining whether certain innate endowments lead a given individual to become the abbot of a new monastic order or the director of a research team—whether medieval mysticism, concern with universal laws of nature, or the urge to develop powerful engines has the most appeal to the gifted mind. As Niels Bohr pointed out in his speech of acceptance of the Atoms for Peace Award, men, like nations, derive their identity and inward quality not so much from the genes they inherit as from the traditions and sense of values imparted by the family in which they are raised and the civilization of which they are a part. Whatever the independence of their behavior, all men are to some extent social parasites who derive their thoughts and preoccupations from their social environment. They are like the statue-man described by Condillac in his famous *Traité des sensations*—a creature which became aware of its needs and developed its ideas in accordance with the environment in which it was placed.

The fact most difficult to explain is that certain places, at certain periods, obviously have provided circumstances peculiarly favorable for the flowering of culture. It is probable that this phenomenon demands some heterogeneity of population to provide the different human types required for any complex cultural system. The social environment must be somewhat unstable to provide opportunities for the emergence of the various human types, yet it must have sufficient permanence to allow time for the development of long-range projects. The concentration of human beings must be adequate—great enough to permit stimulating and fertilizing contacts, not so great as to overpower or dilute intellectual or emotional creativeness. It would be possible to show, I believe, that each great cultural period has been associated with a social structure that incorporated many of these factors, thus permitting the emergence of a creative minority, in the form of a nobility rather than an aristocracy. In an ideal society, selected groups should constantly renew themselves out of the total human substratum. While this general array of conditions seems to be favorable for the emergence of culture, it is obvious that each cultural outburst is determined and characterized by a pattern of determinant causes often difficult to analyze.

Today, in our social structure, technological science is a commodity much in demand, and for this reason educational techniques are being developed for the assembly-line production of the human skills required to manufacture gadgets, products, and cures. Wherever this has been attempted with sufficient vigor, the results have been according to anticipation—most men can become effective technologists if adequately trained. Obviously, man finds it easy to behave as *homo faber* whether his function is to produce pineapples, antibiotics, automobiles, or guided missiles. But *homo sapiens* has never been produced on a large scale, and it is his genesis which is the real puzzle.

Of the many traits recognized among scientists, the most essential—indeed, almost a *sine qua non*—is some form of intellectual curiosity. H. L. Mencken believed that the prototype of the scientist "is not the liberator releasing slaves, the good samaritan lifting up the fallen, but a dog sniffing tremendously at an infinite series of rat holes." This curiosity leads to questions that each individual tries to answer according to his own temperament. There is, in truth, no such thing as a method of discovery. The solution of a problem may come to one man after immense systematic analysis, to another by analogy, to a third as a sudden thought or vision, to yet another as a dream, or in many other ways. There is a method for scientific verification or demonstration, but that is a different thing from discovery. Whichever way discoveries are made, all of them put together constitute the body of scientific knowledge after they have been subjected to proof—either the kind of logical demonstration demanded by the mathematician, or the less convincing verification with which the biologist must be satisfied. Science is made up of the facts and concepts duly demonstrated or verified, then organized into a structure compatible with the philosophical framework socially acceptable at the time.

Granted that intellectual curiosity is one of the most effective forces in the creation of science, the fact remains that the scientist's behavior is determined to a very large extent by factors unrelated to the pursuit of science per se. One of these factors is the reward that intellectual curiosity can bring in the form of popular recognition. Even the withdrawn Darwin was willing to admit in his autobiography: "My love of natural science has been steady and ardent. This pure love has, however, been much aided by the ambition to be esteemed by my fellow naturalists." According to Professor Ludwig Edelstein, the slowness of the development of science in the ancient world can be traced in large part to social indifference. Then, as now, scientists found it necessary to make exaggerated claims, to appeal to mystery, in order to attract public attention. It must be admitted also that the satisfactions of intellectual curiosity are often contaminated with those that come from acquiring a sense of power and domination not only over nature but also over fellow scientists. History provides many examples of these unsavory aspects of scientific life, and so does daily experience in any of our institutions of learning. Conflicts over priority and professional jealousies have always been common at all levels of the scientific population, not barring such heroic figures as Newton and Pasteur.

More difficult to recognize but equally important in the motivation of scientists is the fact that what appears as thirst for knowledge often could be regarded more exactly as a kind of intellectual lust rather than a love of truth. "Most people who call themselves truth seekers do not so much desire to find the truth as to cure their mental itch," wrote Sinclair Lewis in *Arrowsmith*. And, indeed, many investigators have acknowledged that much of the pleasure derived from a discovery or other scientific achievement is of the same order as that experienced in overcoming any difficult situation. "The burning desire for knowledge is what motivates and supports the efforts of the investigator," wrote Claude Bernard; but immediately he added, "The fact that knowledge endlessly recedes as the investigator is about to grasp it, is what constitutes at the same time his torment and his happiness." In a similar spirit Max Planck stated more recently, "It is not the possession of truth, but the success that attends the seeking after it, that enriches the seeker and brings happiness to him." Thus, the pursuit of science has rewards which are independent of the specific nature of its goals and is often akin to an intellectual sport. In this respect again, science does not differ greatly from other human enterprises. Cervantes asserted that "The road is always better than the inn," and Robert Louis Stevenson claimed, in the same spirit, that to travel hopefully is better than to arrive safely.

Curiosity is not sufficient to give intellectual distinction to the activities of scientists. Many animals are inquisitive, some to a high degree. Curiosity acquires scientific and philosophical significance from the nature of the objects to which it is addressed. In this light, Faraday did not do full justice to science if he really said, as recounted in an earlier chapter, that the importance of his discovery of electromagnetic induction was that taxes would eventually be collected from its applications. What he said was true, but not the whole truth.

Faraday was an active member of the very strict Non-conformist sect called Sandemians. His lofty religious ideals and his urge to pursue as far as possible the spiritual implications of his scientific work led him to abandon all industrial consultation work and to renounce the monetary and social advantages that he could have derived from his immense fame. Yet, he does not seem to have ever defended vigorously in public his inner conviction that science is an attempt to understand the universe as much as a technique to exploit nature; that—to use an expression of which he was fond—science is, above and beyond everything else, "natural philosophy." Faraday was not alone in his failure to recognize or at least to acknowledge publicly the philosophical basis of his dedication to science. Pasteur can serve as another example of this attitude, which was apparently very common in the nineteenth century.

It is often claimed that Pasteur's scientific activities originated from a concern with practical problems—for

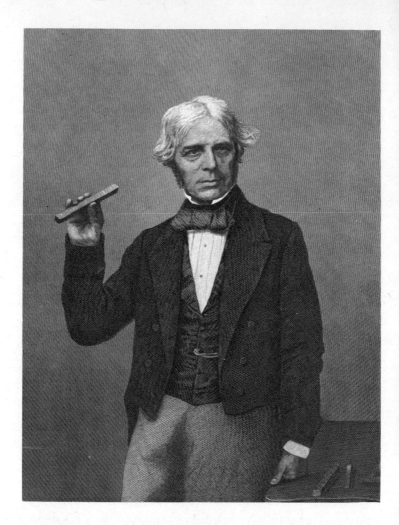

The life of Michael Faraday (1791-1867)
typifies the most elevated aspects of science.
Faraday was a self-educated man, yet his work opened vistas
into the deepest secrets of nature.
A matchless experimenter, he also had
an intuitive perception of eternal truths.
His discoveries constitute the foundation of modern technology,
but for him science was, above everything else, natural philosophy.

example, that his studies on fermentation had their basis in attempts to improve the quality of French wines and beer, or that his interest in infectious processes developed from efforts to save the production of silkworms in France. Nothing could be further from the truth. In reality, as I have shown elsewhere, Pasteur began his scientific life as a purely theoretical investigator and he was already a famous scientist when he began to work on practical problems. From 1847 to 1857 his dominating scientific interests were problems of no apparent practical significance but with large theoretical implications: the relation of molecular structure to optical activity, and the bearing of stereoisomerism on the origin of life; a few years later he became engrossed in other abstract thoughts concerning the biochemical unity of life. As time went on, however, he yielded more and more to the social pressures of his environment, and he spent the largest part of his productive life working on practical problems of fermentation and disease. He became increasingly involved in using science as an instrument of economic conquest rather than as a technique for understanding the universe. Repeatedly he expressed gratification at seeing that his labor would help man to gain mastery over the physical world and to improve human life. "To him who devotes his life to science," Pasteur wrote, "nothing can give more happiness than increasing the number of discoveries, but his cup of joy is full when the results of his studies immediately find practical applications."

There can be no doubt, in my opinion, that Pasteur was aware that his involvement in practical problems had interfered with the pursuit of his deeper scientific interests. He tried to justify his partial neglect of theoretical studies by the statement, "There are not two sciences. There is only science and the applications of science and these two activities are linked as the fruit is to the tree." Yet, despite these brave words and irrespective of his immense success and popular acclaim, it is certain that he often regretted the choice that had been imposed on him by the *Zeitgeist.* Time and time again he stated that he had been "enchained" by an inescapable forward-moving logic that had led him from the study of crystals to the problems of fermentation and then of contagious diseases. He came to believe that it was only through accidental circumstances that he had become involved in practical problems—important, of course, but not so deeply significant as those he had visualized early in life. Yet the desire of his early days to work on crystallography and on the nature of life apparently remained with him as a haunting dream. Pasteur's grandson, Professor L. Pasteur Vallery-Radot, has recently told a moving story which reveals the pathetic intensity of this inner conflict during Pasteur's later years.

I see again that face, that appeared to be carved from a block of granite—that high and large forehead, those grayish-green eyes, with such a deep and kind look. . . .

He seemed to me serious and sad. He was probably sad because of all the things he had dreamed of but not realized.

I remember one evening, at the Pasteur Institute. He was writing quietly at his desk, his head bent on his right hand, in a familiar pose. I was at the corner of the table, not moving or speaking. I had been taught to respect his silences. He stood up and, feeling the need to express his thoughts to the nearest person, even a child, he told me: *"Ah! my boy, I wish I had a new life before me! With how much joy I should like to undertake again my studies on crystals!"* To have given up his research on crystals was the eternal sorrow of his life.

Many modern scientists suffer from the schizophrenic attitude illustrated by the examples of Faraday and Pasteur. Fortunately, one particular aspect of science helps to

Most portraits of Pasteur, either in youth or in old age,
reveal the contemplative aspect of his personality,
his concern with large theoretical problems of science.
Here he is shown in another mood,
dictating to his wife during his practical studies
on silkworm diseases in the South of France.
The violent controversies in which he was then engaged
are reflected in the tenseness of his facial expression.

minimize the inner conflicts generated by this attitude—namely, the fact, already mentioned and universally recognized, that it is often difficult to dissociate the theoretical from the practical aspects of science. Nevertheless, the conflicts are not entirely resolved by this interdependence of theory and practice. The uneasiness of scientists on this score is revealed by the fact that, whereas among themselves they claim that their primary interest is in the conceptual rather than the applied aspects of science, in public they justify basic research by asserting that it always leads to "useful" results, meaning by this the development of processes and products that can be converted into wealth or power. In a symposium on "basic research" recently held in New York, only one of the very distinguished participants dared take the position that the search for knowledge per se is an activity sufficient unto itself, one which does not need further justification.

Like Faraday and Pasteur, other scientists seem to be afraid to admit in public—or do not really believe—that detached intellectual curiosity and desire to understand the universe are proper goals of scientific activity, whether or not "useful" results will ever follow. Yet, despite Francis Bacon's claim that "Knowledge, that tendeth but to satisfaction is but as a courtesan . . . ," the attempt to justify science only by its worldly products is fraught with dangers. Not only does it compromise the intellectual honesty of the scientific community for reasons that need not be discussed here; in my opinion, it also helps to foster among lay people a fundamental skepticism about and even contempt for science itself.

To be scornful of the ultimate intellectual and moral value of natural sciences is, of course, a very ancient attitude. Socrates' skepticism, as expressed in Plato's dialogues, has its counterpart in the talk about the bankruptcy of science that was widespread in literary and philosophical circles at the beginning of the present century. Until recently, however, the attitude of the skeptics was not one of hostility but rather one of impatience and disappointment at the fact that, despite oft repeated promises, science had not yet solved for man the riddle of his nature and his destiny. Far more dangerous, it seems to me, are the expressions of contempt for science as an intellectual discipline, and for scientists as individuals, that have appeared repeatedly during the past few decades. Along with admiration for and awe at the power of science, there exists among the lay public, as pointed out by Margaret Mead, a curious mistrust of the scientist himself, as if he were something scarcely normal and human. This modern attitude toward the scientist is not far removed from that of primitives toward the shaman or medicine man, an individual regarded as essential to the group, but one who is feared and often hated.

As typical of the hostile attitude toward science and scientists, I shall consider two books published respectively in 1913 and 1930: *The Tragic Sense of Life,* by Miguel de Unamuno, and *The Revolt of the Masses,* by José Ortega y Gasset. Both books have been translated into several languages and are still widely read and quoted; they have spread far and wide the doctrine of the bankruptcy of science. Although they deal with different themes, I shall consider them together since they have in common several aspects of the antiscience movement.

Unamuno and Ortega recognize, of course, the contributions made by science to human safety and comfort. But while they appreciate the merits of aspirin and motor cars, they are very little impressed by the kind of intellectual process involved in the technology that has produced these conveniences. Most scientific thinking, according to them, corresponds to a mechanical performance of a rather low order. Just as ancient societies used slaves for the affairs of everyday life, so modern societies produce and use scientific technicians for the same end. Consciously or unconsciously, Unamuno and Ortega have accepted to the letter Bacon's claim that the scientific method is so mechanical and foolproof as to be readily and effectively handled by small minds. They seem to have taken to heart his statement that "brutes by their natural instinct have produced many discoveries, whereas men by discussion and the conclusions of reason have given birth to few or none." As an extension of Bacon's aphorism, it seems worth while to quote at some length from the several pages in *The Revolt of the Masses* that Ortega devotes to the low intellectual caliber of scientists and their discoveries.

"The actual scientific man is the prototype of the mass-man. Not by chance, not through the individual failings of each particular man of science, but because science itself . . . automatically converts him into mass-man, makes of him a primitive, a modern barbarian. Experimental science has progressed thanks in great part to the work of men astoundingly mediocre, and even less than mediocre. That is to say, modern science . . . finds a place for the intellectually commonplace man and allows him to work therein with success. The reason of this lies in what is at the same time the great advantage and the gravest peril of the new science, and of the civilization directed and represented by it, namely, mechanisation. A fair amount of the things that have to be done in physics or in biology is

*This snapshot, taken at 5:30 A.M. on June 16, 1888,
shows Thomas Edison
after seventy-eight hours of continuous work
on his first wax-cylinder phonograph.
Edison's work typifies the use of science
to meet or anticipate social demands,
rather than to answer abstract questions.
Reproduced by courtesy of
the Edison Laboratory National Monument.*

mechanical work of the mind which can be done by any-one, or almost anyone." . . . "The work is done . . . as with a machine, and in order to obtain quite abundant results it is not even necessary to have rigorous notions of their meaning and foundation." "The specialist . . . is not learned, for he is formally ignorant of all that does not enter into his speciality; but neither is he ignorant, because he is 'a scientist,' and 'knows' very well his own tiny portion of the universe. We shall have to say that he is a learned *ignoramus*." "Anyone who wishes can observe the stupidity of thought, judgment, and action shown today in politics, art, religion, and the general problems of life and the world by the 'men of science.'"

Scientists having become so mechanical and unconcerned with philosophical and truly intellectual problems, it is not surprising that, in Unamuno's words, "Science does not satisfy the needs of our heart and our will." Not only does it not deal with the problems of the real man "of flesh and bone," but it "turns against those who refuse to submit to its orthodoxy the weapons of ridicule and contempt."

Thus, according to Unamuno and Ortega, the modern scientist is thoroughly dehumanized, with no horizon beyond his specialized techniques, no awareness of distant human goals. Science fails to deal with the problems that are the real concerns of mankind; and, furthermore, it stultifies all higher aspirations by fostering and satisfying the mass aspects of human nature. Lest there be an illusion that the antiscience movement is peculiar to Latin countries, I shall conclude this discussion with remarks originating from the Anglo-Saxon world. In *The Human Situation,* W. Macneile Dixon asserted that "Science is the view of life where everything human is excluded from the prospect. It is of intention inhuman, supposing, strange as it may seem, that the further we travel from ourselves the nearer we approach the truth, the further from our deepest sympathies, from all we care for, the nearer are we to reality, the stony heart of the scientific universe." As pointed out by Kenneth E. Boulding in *The Image*, many are those all over the world who believe that: "Science might almost be defined as the process of substituting unimportant questions which can be answered for important questions which cannot."

The contempt for science and the scientist illustrated by Unamuno's and Ortega's writings reflects an attitude now fairly widespread. To account for it, scientists are inclined to assume that the public does not have the training or the ability required to appreciate the intellectual beauty and the higher morality of science. But it might be worth while to consider the possibility that the antiscience movement has its origin in the behavior of the scientists themselves—in their own failure to convey to the public the nobler aspects of the scientific heritage, and in misleading assertions which create antagonism. It seems to me, for example, that some scientists have a tendency to derive a kind of unjustified intellectual haughtiness from their familiarity with experimental techniques of which the chief intellectual merit is that they happen to permit the solution of practical problems. These scientists exhibit pride of intellect in speaking of the scientific method as if it were something esoteric, superhuman in its power and precision, whereas it is in reality a very human activity supplemented by a few specialized techniques. Instead of bragging about the purely professional aspects of a "scientific method" that they really cannot define, should not scientists emphasize more than they do the spiritual, creative, and almost artistic aspects of all great scientific advances?

Scientists defend basic research in public by asserting that it cannot fail eventually to yield practical results, but they rarely advertise that knowledge per se is also a precious fruit of science. There is truth, of course, in Benjamin Farrington's statement that "man makes his mental history in the process of conquering the world," but only partial truth. Science, like philosophy, has long been pursued for its own sake, or, rather, for the sake of intellectual satisfaction and increased understanding. Long before science could be justified by its industrial uses Ptolemy experienced the kind of intellectual intoxication that only knowledge can provide. "I know that I am mortal, a creature of a day; but when I search into the multitudinous revolving spirals of the stars, my feet no longer rest on the earth, but, standing by Zeus himself, I take my fill of ambrosia, the food of the gods." In a similar mood Kepler exclaimed, "Eighteen months ago the first dawn rose for me, three months ago the bright day, and a few days ago the full sun of a most wonderful vision." And at the end of his life Pasteur spoke lovingly of "the charm of our studies, the enchantment of science."

It would be interesting to know the reasons which have made such very great scientists as Faraday and Pasteur emphasize the practical worth of their studies and leave unexpressed their loftier intellectual goals. The most obvious interpretation of this attitude is that it was dictated by the wish to gain public approval. But there is no evidence that the public of their time would not have recognized and respected a purely intellectual scientific motivation. In fact, it seems to me that in all situations where the

Bernard de Fontenelle (1657-1757)
initiated the form of science writing
called in French "vulgarisation de la science."
In his hundred years of unmarried life
he brought this technique of writing
to a state of literary perfection,
thus making science a fashionable
topic in the European salons.

public has exhibited any interest in science, it has been just as eager to learn of the philosophical aspects as of the practical applications. True enough, little is known of the manner in which the popularization of science was practiced in past centuries or of the extent of its success; there does not seem to exist any thorough historical study of this interesting aspect of scientific communication. Nevertheless, there readily come to mind the names of many celebrated scientists who achieved great popular acclaim by bringing theoretical knowledge to lay audiences.

Probably the best known of the science popularizers, and certainly one of the first, was not a professional scientist. Bernard de Fontenelle (1657–1757) made his literary reputation with the *Entretiens sur la pluralité des mondes* and continued to hold the limelight with his more austere accounts of the achievements of scientists and of the Académie des Sciences. According to his biographer, L. M. Marsak, Fontenelle's writings had as many readers among the general public of the court and the bourgeoisie as among the learned; they went through six editions in his lifetime, and six more by 1825. Fontenelle mentioned, of course, the practical potentialities of science, but what he emphasized was its humanistic quality, its contribution to enlightenment. He urged on his readers that "Nature is never so admirable or so admired as when it is understood" and that it is at least as important for scientists to help the public rid itself of errors (*"fausses merveilles"*) as to proclaim true marvels. Would that all science reporting respected this admonition in our times! Although it would be out of place to write here at greater length of Fontenelle, I cannot forgo quoting Marsak's appreciation of the magnitude of his achievement. "If the nineteenth and twentieth centuries have held up the businessman and engineer for emulation, it would not be an exaggeration to say that Fontenelle spoke for the civilization of the Enlightenment when he put the scientist in the niche that had formerly been reserved for the artistic creator of the Renaissance or the saint of the high Middle Ages." In addition to educating the public, Fontenelle helped to create a friendly environment that eased the task of the scientists.

At the end of the seventeenth century and throughout the eighteenth century people came to Paris from all over Europe to attend lectures by famous scientists on the theoretical aspects of science. In London the lectures and demonstrations at the Royal Institution long remained a fashionable rendezvous. In Germany, Helmholtz found it worth while to devote much time to presenting various aspects of theoretical science to the general public. Other scientists have on many occasions found responsive audiences eager to learn not of processes and gadgets, but of ideas and general laws. As evidence I need only quote a statement made in 1873 by the physicist John Tyndall in Boston at the end of a highly successful tour during which he had lectured before lay audiences in the United States.

What, I may ask, is the origin of that kindness which drew me from my work in London to address you here, and which, if I permitted it, would send me home a millionaire? Not because I had taught you to make a single cent by science am I here tonight, but because I tried to the best of my ability to present science to the world as an intellectual good. . . . It is specially on this ground of its administering to the higher needs of the intellect; it is mainly because I believe it to be wholesome, not only as a source of knowledge but as a means of discipline, that I urge the claims of science upon your attention.

It will be objected, of course, that times have changed, that the public is no longer interested in the large intellec-

BONES AND STONES, AND SUCH-LIKE THINGS.

Thomas Huxley (1825-1895)
is best known as Darwin's bulldog
in the controversies over evolution,
as a brilliant expositor of nineteenth-century science,
and as an ancestor of Aldous and Julian Huxley.
He deserves also to be remembered
for having contributed indirectly, as one of H. G. Wells's teachers,
to the introduction of a dynamic social attitude
in the modern outlook on science.

tual aspects of science, but is only concerned with what technology can do for human comfort. Although this objection cannot be denied convincingly, a few facts seem to be incompatible with it. For example, many of the books on science for the general public which became best sellers during recent decades dealt not with practical problems, but rather with large theoretical themes of anthropology, biology, physics, astronomy—nay, of mathematics—which could not be in any way practically useful in the conduct of the reader's material life. Here again, an objective study of public response would be enlightening and could provide useful guidance for the popularization of science. Then, it is probably meaningful that, among contemporary scientists, the one best known to the public at large is not one whose name is associated with obvious practical achievements but rather Albert Einstein, who symbolized for the whole world, long before the days of the atomic bomb, the human urge for understanding and for intellectual adventure.

A while ago I used the names of Miguel de Unamuno and Ortega y Gasset to represent the movement which is often called antiscience. This was unfair to these authors because they are, in truth, the voice of humanity begging scientists to remember that man does not live by bread alone. They express also the fear of seeing science identified exclusively with power and technology at a time when it is beginning to reach populations which have never known it under any other guise. It should not be forgotten that in the Western World science was part of culture for several centuries before coming to be used extensively for practical ends. Even today this cultural heritage still conditions to a certain extent the manner in which science is pursued and employed in the countries of Western civilization. In contrast, science is being introduced in the underdeveloped parts of the world not as a cultural pursuit, but merely as a powerful and convenient tool—at best to be used for the production of material wealth, at worst for destructive purposes. For these reasons it seems to me that scientists and science writers betray a public trust when they neglect to emphasize the disinterested aspects of knowledge and are satisfied instead with claiming that all scientific discoveries eventually prove of practical use. On the one hand, there is no evidence that this is true. On the other hand, this attitude ignores the fact that today, as in the past, men starve for understanding almost as much as for food. In the long run the exclusive appeal to practical values may well endanger the future of science and its very existence.

It is obvious, of course, that during recent decades science has improved the lot of man on earth, even more successfully than Francis Bacon had anticipated. It is equally true, however, that for many centuries before the modern era, science had enriched mankind with a wealth of understanding at least as valuable as material riches. Scientists, like other men, win esteem and contribute to happiness more effectively by the exercise of wisdom than by the practice of power. And it is good for them to remember that, long before they had achieved technological mastery over nature and thus become servants of society, their functions as high priests of pure knowledge had given them ancient titles of nobility which they must continue to honor.

7

The Humanness of Science

Nothing could illustrate better the change that occurred in the focus of the scientific community during the Industrial Revolution than the sudden and complete disappearance of the term "natural philosophy." The schism between science and philosophy was the result of two forces which operated almost simultaneously. One was the recognition that knowledge could be used for creating wealth and power; the other was the rapid accumulation of new and unexpected facts which engendered a sense of humility before the complexity of nature and rendered scientists shy of extrapolating from factual knowledge into speculative thoughts. Then humility evolved into scorn for speculation, and today the statement "This is not science, this is philosophy" rules out of scientific discussion any statement that goes a step beyond established fact.

Yet it is apparent that today, as in the past, many scientists—among them some of the most brilliant and most effective—are eager to escape from the austere discipline of factual knowledge and to experience again the intoxication of philosophical thought. They may distrust Plato, but, like him, they seem to regard philosophy as the "dear delight." Witness the flurry of speculative books published by scientists as soon as some discovery enlarges the scope of their knowledge. The theory of evolution has been used by biologists as a platform to erect or justify religious, political, and economic philosophies. Familiarity with modern theoretical physics seems to warrant opinions not only on the structure of matter and its relation to energy, but also on the nature of life, the existence of free will, or the symbolism of language.

This return to scientific philosophy negates, it seems to me, the fears so commonly expressed that scientists are becoming a class apart from the rest of society by developing a culture without contact with the rest of human life. It is true, of course, that within the area of his particular work each scientist becomes so specialized that he finds it difficult to communicate on scientific subjects except with other workers in the same specialized field. But this situation is not peculiar to science. It exists just as much in other forms of learning—in philology or Moslem culture as much as in mathematics or genetics. Moreover, science should not be regarded as one single discipline concerning which sweeping statements can be made, any more than this can be done for the so-called humanities. With regard to the knowledge and operations defined by their techniques, the biologist and the mathematician are as far apart as they are from the student of Sanskrit or from the art critic, and as these are from each other. We must accept as a fact that the modern world is made up of an immense number of specialized groups, intellectually separated by experience, words, and the meaning of symbols. But although the members of one intellectual guild can rarely understand the professional jargon of another guild, men can and do communicate nevertheless at a higher level of discourse. Experiences, words, and symbols can usually be reformulated in the context of broader human meanings. In my opinion, there are not "two cultures," even though C. P. Snow has made the expression famous. There are a multiplicity of intellectual occupations, each of which fortunately has several points of contact with human life. Whatever his field of specialization, the scholar can be understood beyond the confines of his guild—but only if he is willing to raise his language above the jargon of his trade. The scholar must learn to speak to man.

Considered from the point of view of this higher level of discourse, natural sciences do not seem to be more esoteric, or more "dehumanized," than are other fields of learning. Indeed, it is questionable whether the so-called "humanities" are intrinsically more meaningful and have greater interest for the general public than scientific studies. Literature, music, and the plastic arts owe their popular appeal not to their intellectual content, but to their emotional quality, to their play on love, jealousy, hatred, and other passions. When they are reduced to intellectual and technical presentations, humanistic studies are just as devoid of popular appeal as are scientific studies. As to science, it becomes a popular topic whenever it deals with a subject loaded with emotional value, whether it be the origin of man, the conquest of space, or the sexual norms of the American male.

It is doubtful that there exists a definition of culture which is universally acceptable. Yet in ordinary social contacts it is not difficult to identify types of behavior which almost everybody would accept as corresponding to a cultured way of life. For the scientist a cultured attitude implies the ability and willingness to relate his field of work to historical developments, to emphasize its bearing on the future, and, more generally, to recognize its relevance to other human interests. It demands an awareness that science is a humanistic activity to the extent that it is more than a body of facts and techniques and that it deals with material meaningful to the preoccupations of mankind. There are, of course, many specialists who are not concerned with the cultural, humanistic implications of science, and who are satisfied with the artisan aspects of their professions, but this is true in all fields of scholarship. As far as I can judge, science meets all the requirements usually associated with the concepts of culture and humanism. Furthermore, I see no reason to believe that the scientific way of life is any less compatible than others with the cultivation of a cultured attitude. The learned ignoramus that Ortega y Gasset found among scientists is just as familiar a type in the humanities as in science.

More positively, it is certain that science is concerned with universals, with an approach to truth, with the recognition of patterns and the creation of concepts, and, last but not least, with the perception of sensuous as well as of abstract beauty. As these statements are somewhat outside my theme, I shall not develop them here. Instead, I shall return briefly to some of the topics discussed in the preceding chapters in order to illustrate that modern science is maintained in the channel of general culture by the need to reexamine constantly the meaning and relevance of its social and philosophical assumptions.

As we have seen, the transformations of human life which have taken place during the past hundred years are the realizations of the utopias formulated by the seventeenth- and eighteenth-century philosophers. Not so long ago the role of the scientist in this enterprise appeared straightforward and all to the good; each advance in scientific knowledge eventually resulted in some contribution to human health and happiness. Confident of the ultimate beneficence of his work, the scientist had good reasons to keep aloof from social problems. It is obvious, however, that the situation is now changing rapidly, and one can anticipate that the scientist will face more and more problems of conscience as the social power of science continues to increase. The necessity for the scientist to reexamine his activities in the light of social considerations can be illustrated by quotations from the recent report of a committee appointed by the American Association for the Advancement of Science to consider "Science and Human Welfare." In the words of the report:

Science is being consciously exploited for industrial, military, and political purposes. At the same time there is little recognition of the internal needs of science, or of its purposes as a discipline of the human mind. . . . Having become a major instrument in political affairs, science is inseparably bound up with many troublesome questions of public policy. That science is valued more for these uses than for its fundamental purpose—the free inquiry into nature—leads to pressures which have begun to threaten the integrity of science itself.

When the social implications of science are discussed, the issues that immediately come to mind are certain obvious threats to mankind such as those associated with atomic power or with population pressures. In fact, however, the scientist's responsibility is involved in many other issues which appear less dramatic, perhaps, but are probably as important in the long run and more difficult to solve because less clearly defined. Until very recent times so little could be done to deal with the obvious shortages and sufferings in the world that the most urgent need was to develop techniques for the production of material wealth and for the control of disease. Now the power of science is so great that almost any desired method, gadget, or product can be developed if we are willing to devote enough resources to the task. And it is precisely the confidence that utopias can now be converted into realities which creates urgent ethical problems for the scientist.

The question of how to do things was a purely technical one that could be decided on scientific criteria; but the choice of *what* to do, among all the things than can be done, clearly implies some concern with ultimate social consequences. There is no longer any thoughtful person who believes that the conversion of science into more power, more wealth, or more drugs necessarily adds to health and happiness or improves the human condition. Indeed, haphazard scientific technology pursued without regard for its relevance to the meaning of human life could spell the end of civilization. Unless he becomes concerned with social philosophy, the scientist will increasingly hear the words of Oscar Wilde applied to him: that he knows the price of everything, but the value of nothing.

While the purely theoretical scientist is not directly concerned with problems of social philosophy, he faces other philosophical problems in having to redefine the conceptual basis of his knowledge. Discussions of causality, of the uncertainty principle, of the impossibility of dissociating the observer from the observed event are now equally meaningful for physics and for philosophy. Similarly, it is difficult to believe that biology can long retain its quaint nineteenth-century provincialism and delay reexamining the validity of the Cartesian assumptions under which it operates at present, such as the mind-matter duality and the body-machine concept. The fact that Bergson's *élan vital* did not constitute a positive contribution to biological theory is no excuse for ignoring his criticism in *Creative Evolution* that "we treat the living like the lifeless. . . . We are at ease only in the discontinuous, in the immobile, in the dead. The intellect seems to be characterized by a natural inability to comprehend life."

It would seem unnecessary to mention here the processes through which new facts are discovered and their validity and significance are tested. No one any longer assumes that science advances through a mechanical application of the Baconian inductive method or of Descartes's deductive rationalism. On the other hand, it is useful to be clearly aware that much of scientific knowledge is concerned not with objects and events as they occur naturally, but with fragments of nature produced artificially by analytical breakdown. Indeed, this analytical approach to knowledge is perhaps the one characteristic that most sharply differentiates science from art in the attempt to apprehend reality, and there seems to be an increasing tendency to identify the "scientific method" with this approach. In reality, however, some of the great leaps in the scientific advance have come not from the detailed analysis of mechanisms but rather from the total "intuitive" apprehension of reality. Michael Faraday is one of the famous scientists whose discoveries and writings strikingly illustrate the role of intuition in science. Faraday was an unfatigable and a peerless experimenter, but in addition he had a kind of divinatory gift that made him perceive, for example, the existence of the fields of forces which were only later described in mathematical terms.

At the end of a series of experiments that he had hoped would establish a relation between magnetism and other phenomena, he was sincere enough to confide to his notebook, "The results are negative. They do not shake my strong feeling of the existence of a relation between gravity and electricity, though they give no proof that such a relation exists." Elsewhere he stated, "I have long held an opinion, almost amounting to conviction . . . that the various forms under which the forces of matter are made manifest have one common origin; . . . that they are convertible, as it were, one into another."

Those who knew Faraday were aware of the visionary, almost prophetic aspect of his scientific personality. *"Er riecht die Wahrheit"* (He smells the truth) Kuhlraush said of him. John Tyndall spoke of his "flashes of wondrous insight and utterances which seem less the product of reasoning than of revelation." And Clerk Maxwell obviously was comparing Faraday's concrete vision with his own abstract analytical mind when he wrote that scientific truth is equally valid whether it appears "in the robust form and the vivid colouring of a physical illustration, or in the tenuity and paleness of a symbolical expression."

While emphasizing the limitations imposed on Faraday by the inadequacy of his theoretical training, Professor Charles C. Gillispie also recognized in his book *The Edge of Objectivity* that "there was given to him as to few scientists a sense of the spatial. He would almost see the moving wire slice through the lines of force and the current stir within. Perhaps, after all, it was the reward of his incapacity for abstraction, this *vision of nature in the round, and in depth* . . ." (Italics mine.) It is perhaps this "vision of Nature in the round" that Einstein had in mind when he wrote in his autobiography, "Physics is an attempt to grasp reality as it is thought, independently of it being observed."

Granted that certain individuals with special gifts have been able to advance knowledge by a kind of intuition, it must be admitted nevertheless that the most concrete and also the most characteristic achievements of Western thought have been the products of analytic rationalism. But it is possible that as a result of cultivating with such intensity the abstract, conceptual aspects of science, we have allowed our endowments for perception of nature in the round to become atrophied, or at least blunted. Although the attribute of perception cannot be measured or defined, experience shows that persons differ in the degree to which they can apprehend as a whole the multiple aspects of reality and the interrelations between them. As Aldous Huxley stated in *The Doors of Perception,* "We must preserve and if necessary intensify our ability to look at the world directly and not through the half-opaque medium of concepts, which distorts every given fact into the all too familiar likeness of some generic label or explanatory abstraction."

Needless to say, reintroducing "perception" as a technique in science involves the danger of substituting meaningless generalities for the concreteness of exact knowledge, of drowning facts and laws in the morass of loose words and vague concepts. But insistence on clearness of thought and of purpose need not deter us from recognizing that in the long run knowledge might be enriched by cultivating the awareness that almost everything is relevant to everything else. Reality has multiple facets and, therefore, can be apprehended only if seen from different points of view. The concept of complementarity is, of course, a manifestation of this awareness.

One of the intellectual steps that led to modern science was the recognition that man cannot discover the laws of the universe, let alone achieve mastery over nature, by the mere exercise of reason. Experimentation is the safest and most rapid road to knowledge and to power. But the fact that scientific knowledge is now gained mostly by experimentation does not decrease the power and the dangers of reason. What the experimenter does, and consequently what he finds, is determined to a large extent by his assumptions. Thus, in practice, the facts revealed to the experimenter are limited by the range of his reason, and not uncommonly they are distorted by the very dreams of his reason. To a large extent also, the experimenter is like Antaeus, in Greek mythology, whose strength ebbed as soon as he lost contact with the earth and could be renewed only when he again touched the ground. The perceptions of the scientist are the contacts through which his creative power is kept alive and healthy. They provide the substantial food which generates a deeper kind of objectivity, more true fundamentally than factual evidence. Deprived of this human quality, reason tends to spin cobwebs, or to create nightmares that alienate science from man.

Individual scientists may display an intellectual arrogance and lack of humanism fostered by confidence that their knowledge means wealth and power. As a class, however, scientists have remained humble. They realize with Democritus that "We know nothing unerringly but only as it changes according to the disposition of our body and the things that enter it and impinge upon it." Fortunately, limitation of understanding is compatible with a creative social attitude. In fact, the dynamic character of modern science is due in large part to the manner in which it has put into practice Kant's definition of faith: namely, holding a concept to be true on grounds that are adequate for action, although they may not be sufficient to satisfy the intellect.

Above and beyond these pragmatic considerations, scientists are also convinced that what they do has a bearing on the apprehension of absolute truth. Whatever questions are asked concerning the universe and human destiny, the answers of theologians and of philosophers must be consistent with the demands of informed intelligence—that is, with scientific facts. No one can question, furthermore, that the world of living as well as of lifeless things revealed by scientific investigation is incomparably grander than anything that emerges from abstract thought or from the most vivid imagination. By uncovering some of the hidden mechanisms of appearances, science opens to contemplation and meditation new, unexpected vistas. Science is like a revelation that enlarges awareness by sharpening and extending the direct perceptions from which philosophy originated.